TIME LIFE® BOOKS

Other Publications:

PLANET EARTH
COLLECTOR'S LIBRARY OF THE CIVIL WAR
LIBRARY OF HEALTH
CLASSICS OF THE OLD WEST
THE EPIC OF FLIGHT
THE SEAFARERS
THE ENCYCLOPEDIA OF COLLECTIBLES
THE GREAT CITIES
WORLD WAR II
HOME REPAIR AND IMPROVEMENT
THE WORLD'S WILD PLACES
THE TIME-LIFE LIBRARY OF BOATING
HUMAN BEHAVIOR
THE ART OF SEWING
THE OLD WEST
THE EMERGENCE OF MAN
THE AMERICAN WILDERNESS
THE TIME-LIFE ENCYCLOPEDIA OF GARDENING
LIFE LIBRARY OF PHOTOGRAPHY
THIS FABULOUS CENTURY
FOODS OF THE WORLD
TIME-LIFE LIBRARY OF AMERICA
TIME-LIFE LIBRARY OF ART
GREAT AGES OF MAN
LIFE SCIENCE LIBRARY
THE LIFE HISTORY OF THE UNITED STATES
TIME READING PROGRAM
LIFE NATURE LIBRARY
LIFE WORLD LIBRARY
FAMILY LIBRARY:
   HOW THINGS WORK IN YOUR HOME
   THE TIME-LIFE BOOK OF THE FAMILY CAR
   THE TIME-LIFE FAMILY LEGAL GUIDE
   THE TIME-LIFE BOOK OF FAMILY FINANCE

*This volume is one of a series that explains and demonstrates
how to prepare various types of food, and that offers in each
book an international anthology of great recipes.*

Dried Beans & Grains

BY
THE EDITORS OF TIME-LIFE BOOKS

TIME-LIFE BOOKS/ALEXANDRIA, VIRGINIA

Cover: Boston baked beans, drenched in molasses-sweetened sauce, are lifted from their pot. They were cooked with salt pork, onions and dry mustard (pages 70-71; recipe, page 105).

Time-Life Books Inc.
is a wholly owned subsidiary of
TIME INCORPORATED

Founder: Henry R. Luce 1898-1967

Editor-in-Chief: Henry Anatole Grunwald
President: J. Richard Munro
Chairman of the Board: Ralph P. Davidson
Executive Vice President: Clifford J. Grum
Chairman, Executive Committee: James R. Shepley
Editorial Director: Ralph Graves
Group Vice President, Books: Joan D. Manley
Vice Chairman: Arthur Temple

TIME-LIFE BOOKS INC.

Managing Editor: Jerry Korn. *Text Director:* George Constable. *Board of Editors:* Dale M. Brown, George G. Daniels, Thomas H. Flaherty Jr., Martin Mann, Philip W. Payne, Gerry Schremp, Gerald Simons. *Planning Director:* Edward Brash. *Art Director:* Tom Suzuki; *Assistant:* Arnold C. Holeywell. *Director of Administration:* David L. Harrison. *Director of Operations:* Gennaro C. Esposito. *Director of Research:* Carolyn L. Sackett; *Assistant:* Phyllis K. Wise. *Director of Photography:* Dolores A. Littles. *Production Director:* Feliciano Madrid; *Assistants:* Peter A. Inchauteguiz, Karen A. Meyerson. *Copy Processing:* Gordon E. Buck. *Quality Control Director:* Robert L. Young; *Assistant:* James J. Cox; *Associates:* Daniel J. McSweeney, Michael G. Wight. *Art Coordinator:* Anne B. Landry. *Copy Room Director:* Susan B. Galloway; *Assistants:* Celia Beattie, Ricki Tarlow

Chairman: John D. McSweeney. *President:* Carl G. Jaeger. *Executive Vice Presidents:* John Steven Maxwell, David J. Walsh. *Vice Presidents:* George Artandi, Stephen L. Bair, Peter G. Barnes, Nicholas Benton, John L. Canova, Beatrice T. Dobie, Carol Flaumenhaft, James L. Mercer, Herbert Sorkin, Paul R. Stewart

THE GOOD COOK

The original version of this book was created in London for Time-Life International (Nederland) B.V.
European Editor: Kit van Tulleken; *Design Director:* Louis Klein; *Photography Director:* Pamela Marke; *Planning Director:* Alan Lothian; *Chief of Research:* Vanessa Kramer; *Chief Sub-Editor:* Ilse Gray; *Production Editor:* Ellen Brush; *Quality Control:* Douglas Whitworth

Staff for *Dried Beans & Grains:* Series Coordinator: Liz Timothy; *Head Designer:* Rick Bowring; *Text Editor:* Gillian Boucher; *Anthology Editors:* Josephine Bacon, Liz Clasen; *Staff Writer:* Thom Henvey; *Researchers:* Krystyna Davidson, Margaret Hall, Deborah Litton; *Sub-Editors:* Katie Lloyd, Sally Rowland; *Design Assistants:* Cherry Doyle, Adrian Saunders; *Editorial Department:* Pat Boag, Kate Cann, Debra Dick, Beverley Doe, Philip Garner, Molly Sutherland, Julia West, Helen Whitehorn

U.S. Staff for *Dried Beans & Grains:* Editor: Gerry Schremp; *Senior Editor:* Ellen Phillips; *Designer:* Ellen Robling; *Chief Researcher:* Barbara Fleming; *Picture Editor:* Christine Schuyler; *Writers:* Patricia Fanning, Leslie Marshall; *Researchers:* Mariana B. Tait (techniques), Robert Carmack (anthology), Denise Li, Ann Ready; *Assistant Designer:* Peg Schreiber; *Copy Coordinators:* Nancy Berman, Tonna Gibert, Bobbie C. Paradise; *Art Assistants:* Robert Herndon, Mary L. Orr; *Picture Coordinator:* Alvin Ferrell; *Editorial Assistants:* Brenda Harwell, Patricia Whiteford

CHIEF SERIES CONSULTANT

Richard Olney, an American, has lived and worked for some three decades in France, where he is highly regarded as an authority on food and wine. Author of *The French Menu Cookbook* and of the award-winning *Simple French Food,* he has also contributed to numerous gastronomic magazines in France and the United States, including the influential journals *Cuisine et Vins de France* and *La Revue du Vin de France.* He is a member of several distinguished gastronomic societies, including L'Académie Internationale du Vin, La Confrérie des Chevaliers du Tastevin and La Commanderie du Bontemps de Médoc et des Graves. Working in London with the series editorial staff, he has been basically responsible for the planning of this volume, and has supervised the final selection of recipes submitted by other consultants. The United States edition of The Good Cook has been revised by the Editors of Time-Life Books to bring it into complete accord with American customs and usage.

CHIEF AMERICAN CONSULTANT

Carol Cutler is the author of a number of cookbooks, including the award-winning *The Six-Minute Soufflé and Other Culinary Delights.* During the 12 years she lived in France, she studied at the Cordon Bleu and the École des Trois Gourmandes, and with private chefs. She is a member of the Cercle des Gourmettes, a long-established French food society limited to just 50 members, and is also a charter member of Les Dames d'Escoffier, Washington Chapter.

SPECIAL CONSULTANT

Jeremiah Tower is an eminent American restaurateur who lived for many years in Europe, and is a member of La Commanderie du Bontemps de Médoc et des Graves and La Jurade de Saint-Émilion. He has been largely responsible for the step-by-step photographic sequences in this volume.

PHOTOGRAPHER

Aldo Tutino, a native of Italy, has worked in Milan, New York City and Washington, D.C. He has won a number of awards for his photographs from the New York Advertising Club.

INTERNATIONAL CONSULTANTS

GREAT BRITAIN: *Jane Grigson* has written a number of books about food and has been a cookery correspondent for the London *Observer* since 1968. *Alan Davidson,* a former member of the British Diplomatic Service, is the author of several cookbooks and the founder of Prospect Books, which specializes in scholarly publications about food and cookery. *Jean Reynolds,* who prepared some of the dishes for the photographs in this volume, is from San Francisco. She trained as a cook in the kitchens of several of France's great restaurants. FRANCE: *Michel Lemonnier,* the cofounder and vice president of Les Amitiés Gastronomiques Internationales, is a frequent lecturer on wine and vineyards. GERMANY: *Jochen Kuchenbecker* trained as a chef, but worked for 10 years as a food photographer in several European countries before opening his own restaurant in Hamburg. *Anne Brakemeier* is the co-author of a number of cookbooks. ITALY: *Massimo Alberini* is a well-known food writer and journalist, with a particular interest in culinary history. His many books include *Storia del Pranzo all'Italiana, 4000 Anni a Tavola* and *100 Ricette Storiche.* THE NETHERLANDS: *Hugh Jans* has published cookbooks and his recipes appear in several Dutch magazines. THE UNITED STATES: *Judith Olney,* author of *Comforting Food* and *Summer Food,* received her culinary training in England and France. In addition to conducting cooking classes, she regularly contributes articles to gastronomic magazines.

Correspondents: Elisabeth Kraemer (Bonn); Margot Hapgood, Dorothy Bacon (London); Susan Jonas, Lucy T. Voulgaris (New York); Maria Vincenza Aloisi, Josephine du Brusle (Paris); Ann Natanson (Rome).
Valuable assistance was also provided by: Karin B. Pierce (London); Bona Schmid, Maria Teresa Marenco (Milan); Miriam Hsia, Christina Lieberman (New York); Michèle le Baube (Paris); Mimi Murphy (Rome).

First printing. Printed in U.S.A.
Published simultaneously in Canada.
School and library distribution by Silver Burdett Company, Morristown, New Jersey 07960.

TIME-LIFE is a trademark of Time Incorporated U.S.A.

For information about any Time-Life book, please write:
Reader Information, Time-Life Books
541 North Fairbanks Court, Chicago, Illinois 60611

Library of Congress CIP data, page 176.

CONTENTS

A Culinary Voyage of Discovery

Dried beans and their culinary cousins, dried lentils and peas, are found in practically every kitchen cupboard, and the same is true of such grains as rice and oats. Yet these venerable foods too often are treated with little imagination, simply because they seem so ordinary that many cooks overlook their possibilities. First among these is the sheer diversity of the elements: Dried beans, lentils and peas appear in varieties by the score; as for grains, a dozen or more types are grown worldwide, each processed in a broad range of forms—whole, cracked, rolled and ground. Then, although boiling is almost always fundamental to their preparation, even subtle changes and elaborations of cooking technique can give strikingly different results. Above all, these staples combine gracefully and happily with many other ingredients.

The first half of this book is designed not only to teach the principles of preparing beans and grains, but also to offer a survey of the frequently surprising ways in which they are presented in countries as widely separated as China, France, Iran and Brazil. On the pages that follow, you will find a culinary history of these staples, as well as illustrated guides that display their multiplicity. The succeeding three chapters demonstrate how various cooking methods—boiling, steaming, frying and baking—are applied to beans and grains; a final chapter gives instructions for the construction of such classic assemblies as the pilafs of India and paellas of Spain. The second half of the volume consists of an international anthology of more than 200 recipes from 41 nations.

The elemental pulses

Dried beans, lentils and peas are known collectively as beans or as pulses (the latter term derives from the Latin *puls,* a primitive grain or bean porridge). All are seeds of leguminous plants, and they either have been allowed to dry in their pods before being shelled or have been shelled and then dried by mechanically circulated air.

Leguminous plants flourish untended in many different environments, yet respond well to cultivation. Their yield can be dramatically increased with relatively little effort. Many even enrich the soil in which they grow. Once harvested, their dried seeds keep for years. And when rehydrated—by simply cooking them in liquid—beans double or triple in size to become a satisfyingly bulky and nutritious food: Their average protein content is 22 per cent by weight, while the soybean—sometimes called the "cow of the East"—is 40 per cent protein.

For these reasons, it is not surprising that beans were among the earliest plants domesticated. Lentils, for example, have been found by archeologists in Middle Eastern sites dating back nearly 9,000 years. And the large family of common beans that includes the kidney, navy and pinto beans was farmed in Central America by 5000 B.C. Soybean cultivation was well advanced in China during the Shang Dynasty almost 4,000 years ago. By then, fava beans were being farmed in the Near East and North Africa; favas were found in the ruins of Troy.

To envisage the adaptable nature of beans, consider the spread and shifts of their centers of cultivation over the centuries. For instance, when Columbus arrived in the New World, the common bean that had originated in tropical Mexico nourished Indians in the cold northern regions of North America as well as in its hot, dry Southwest, and was widely farmed in mountainous Peru as well as on the plateaus of Chile. Introduced to Europe in the 16th Century, the common bean quickly supplanted the fava—then Europe's primary bean.

The peregrinations of other beans span the globe in similar fashion. Chick-peas, originally from Southwest Asia, were taken to Sicily and Switzerland in prehistoric times, then traveled throughout the countries around the rim of the Mediterranean, east to India, and west to Brazil and Mexico. As for lentils, they are now an important crop in India, Spain, Chile and Argentina.

A variety of presentations

Dried beans have traditionally had a reputation as peasant fare, a reflection of their central role in the winter diet of poor folk the world over. Because of these humble associations, sophisticated diners have tended to be dismissive. But dried beans possess a special culinary virtue: They are almost invariably cooked in liquid and, in the process, these bland seeds can absorb complex and subtle flavors. A spectacular reminder of this quality is *dhansak,* a specialty of the Parsi sect of western India. A stew of lamb, tripe and vegetables, *dhansak* may contain anywhere from three to nine varieties of lentils, chosen so as to produce a refined blend of tastes when simmered with herbs and spices such as ginger, garlic, cloves, cinnamon, coriander and mint.

Stews of this type—rich with fatty meats, bright with flavorings—appear in many lands. Americans esteem the baked beans of New England, fragrant with molasses and rich with salt pork—like the beans, a winter staple *(pages 70-71).* The many versions of France's cassoulet *(pages 74-77)* offer even more elaborate blends of meat and beans, as does Brazil's hearty *feijoada completa (pages 78-81).*

There are, however, numerous bean dishes that preserve a

less complicated, rustic appeal. Among them are puréed seasoned beans served as hors d'oeuvre or as side dishes. India, for example, offers a wide range of fragrant, full-bodied or soupy lentil *dals (recipes, pages 18-19);* Italian cooks make olive oil-flavored purées of white kidney beans. In Mexico, pinto or red kidney beans are transformed into lightly crusted *frijoles refritos*—well-fried beans *(pages 44-45).*

Nor is delicacy out of the question: As Asian cooks long have known, liquid extracted from soybeans can be turned into silky bean curd *(pages 20-21).* Puréed cooked beans, enriched with cream and lightened with eggs, become agreeable custards *(pages 64-65).* And any sort of whole bean, pea or lentil can be served with no more garnish than a little butter or oil. A notable example of such artful simplicity is the Florentine dish, *fagioli al fiasco* (beans in a flask), made by simmering beans gently in a bottle whose narrow neck prevents the escape of steam and flavor, then serving the beans drizzled with olive oil.

The life-sustaining grasses

Unlike beans, grains are accorded high honors in the culinary world, and rightly so: Grains are the mainstay of human sustenance. Approximately half of the world's arable land is devoted to the cultivation of grain—and the plants provide 80 per cent of the calories that humanity consumes. Their lineage is at least as ancient as the lineage of beans. By 7000 B.C., the cultivation of the grasses that yield wheat and similar grains had transformed nomadic Middle Eastern peoples, who once lived an uncertain and unpredictable life by hunting animals and gathering plants, into secure and stable communities, nourished by the fruits of their own agriculture.

It may seem strange that weedlike plants such as grasses were so important in ushering in the era of agriculture. In fact, the explanation probably lies in the grasses' weedlike qualities—their ability to spring up from any odd patch of bare ground on which a seed had happened to fall. And grasses have many other characteristics that were—and still are—extremely valuable in a crop. They take up little room, sending up stems topped by crowded spikes of nutrition-packed seed kernels. They mature in just a few months, and all of the seeds ripen simultaneously. They are easily prepared for cooking, even with primitive equipment—a sickle to cut the stalks, a stick to loosen the individual kernels, and some sort of pestle and mortar to remove the inedible husk that encases each kernel. Best of all, grains are often dry enough when fully ripe or after brief parching in the sun to be stored without becoming moldy; thus, a good harvest ensures a year-round supply of food. From the taming of wheat, there emerged—over the ages—the great ancient civilizations of Sumer and Egypt, and later of Greece and Rome.

Ideally suited to temperate, fairly dry Mediterranean lands, wheat did less well in chillier, wetter regions. But oats, rye and buckwheat—this last not a grain but a plant with a similar, somewhat softer seed—could withstand the harsher climates. These plants had grown among the wheat and were eventually cultivated for their own sakes wherever wheat was a less successful crop. Now their cultivation is widespread, although rye and buckwheat are more prominent in Eastern Europe and the

Soviet Union, and oats are particularly esteemed in Scotland.

Millet—also important to ancient civilizations—is one of the hardiest of grains; it will grow in arid climates where no other grain can survive. Although little known except as birdseed in the United States and Canada, the hard-hulled millet is a staple in the desert regions of Asia and Africa, and its mild flavor is well worth sampling.

Rice is as important a crop as wheat, but requires quite different growing conditions and, therefore, was first cultivated in quite different areas. Rice is a grass that grows best when half-submerged with water—in paddy fields. It was first domesticated in Asia. Cultivated rice kernels reportedly dating from 3500 B.C. have been found in eastern Thailand, and by 3000 B.C., rice farming in India was well advanced.

Caravans and ships trading chiefly in precious silks and spices also brought seed rice to the Near East and Africa. But the grain did not appear in Europe until the Moors invaded Spain in the Eighth Century A.D. Cultivation then spread into Italy and, in the late Middle Ages, into southern France. In the United States, rice was first grown successfully in South Carolina in the 1690s; it did so well that it came to be called "Carolina gold." However, cotton replaced it in South Carolina, and the U.S. rice crop is now grown primarily in Arkansas, Louisiana, Texas and California.

North America had its own native grain that, like rice, grew in marshy land—mostly around the western perimeter of Lake Superior. French explorers dubbed it "crazy oats" and English

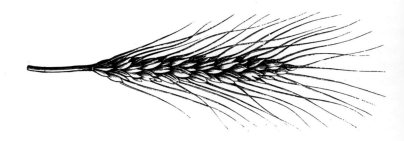

settlers named it "wild rice," although, in fact, the plant is unrelated to either oats or rice. Until about 20 years ago, the grain remained wild, and the great majority of it was harvested in Minnesota by Chippewa Indians, who glided among the plants in canoes and flailed the stalks until the seeds fell into their crafts. Today, almost half of all wild rice is farmed in paddy fields—in California as well as in Minnesota and Wisconsin. However, because it is still difficult to produce in large quantities, wild rice remains extremely expensive: It could be categorized as the truffle of the grain world. And, like truffles, it is decidedly worth the expense: Its full, nutty flavor compares with that of no other grain.

Another grass native to the New World is maize, or, as it is now called in the United States and Canada, corn. Americans are most familiar with sweet corn, which they eat fresh; field corn—the bland-flavored and oldest type of the species—is now widely farmed for fodder, but its original use was as food for man. Even today field corn, dried and processed in various ways, yields hominy, hominy grits, cornmeal and even flour. Maize — the "seed of seeds" to prehistoric Indian farmers — was cultivated almost 7,500 years ago in Central America and soon spread throughout North and South America; it made possible the glittering Mayan, Aztec and Inca civilizations. As the Spanish conquistadors of the 16th Century noted, the Aztec emperor Montezuma and his compatriots ate a wide variety of corn dishes, including tamales; these, demonstrated on pages 36-37, are prepared much today as they were then.

Seed corn was carried to Spain by the conquistadors, and from there its cultivation spread to other Mediterranean areas. Although the only type widely grown in Europe is field corn, the northern Italians have adopted it as a basic grain. Dried and ground, corn provides them with polenta, the modern and delicious descendant of the porridges of ancient times.

The cooking of grains

Among grains, rice would seem to be the one with the greatest culinary potential. Writing in the early years of this century, the French chef Auguste Escoffier pronounced firmly that "rice is the best, the most nutritive and unquestionably the most widespread staple in the world." Escoffier's statement was only partly true; rice in its most common, polished form is less nutritious than whole grains that still retain their protein-laden germs. However, white rice is enjoyed by far more people than other whole grains and lends itself to a generous range of treatments. Long-grain rice, cooked in butter or oil and simmered with herbs, spices, vegetables and meats, provides the variegated pilafs of India and the Middle East (pages 24-25 and 82-83). Short-grain rice, cooked in a related manner and with a similar spectrum of ingredients, produces Italy's creamy risottos, studded with bright garnishes (pages 26-27). And rice cooked by the basic methods of boiling or simmering can serve as the foundation of any number of elaborations. It may be deep fried to make golden-crusted croquettes (pages 56-57), for instance, or molded into a shell to hold rich meaty fillings (pages 62-63).

Rice is by no means the only grain that offers a tempting range of presentations. Although wheat and rye are most often ground to flour for bread, and barley's chief employment is in brewing beer, all of these grains — and many others — can be used either whole or in their cracked forms in the same manner as rice to render succulent kernels that have a range of flavors from nutty to sweet.

Ground into grits or meals, grains open yet another avenue for exploration. Like other forms of grain, grits and meals must be cooked in liquid. The simplest results of this cooking are homely breakfast or dessert cereals — cornmeal mush, oatmeal porridge and molasses-scented Indian pudding (pages 68-69). However, these porridges have a peculiar characteristic: When spread out and cooled, they harden into firm sheets that can then be cut into shapes and fried (pages 46-47) or baked (pages 60-61) to produce crisp-surfaced but delectably absorbent platforms for sauces and garnishes. Nowhere is this technique better understood than in Italy, whose cooks for centuries have turned polenta into the foundation of innumerable rich presentations (recipes, pages 150-151). Polenta can be served with cheese, tomato or meat sauces, with stewed vegetables and even with sugary sauces. And its somewhat sweet taste is the perfect companion to roasted game in presentations such as that lovingly described by the 19th Century Italian poet Lorenzo Stecchetti: "The thrushes / More than thirty / In superb majesty / Were seated on the polenta / Like Turks on a divan."

Serving bean and grain dishes

Dishes founded on dried beans, lentils, peas or grains can serve any of a number of different functions in a meal; the choice depends on how the dishes are made and with what complementary ingredients. Such delicate collations as the small bean custards shown on page 58 would make a delicious first course; in

Italy, small portions of risotto often play the same role. Many simple rice or bean dishes serve as accompaniments to a main course, but hearty, meaty stews — cassoulet and pilaf, for instance — are main courses in themselves. Some elaborate assemblies such as paella or *feijoada* are complete meals. And sweet bean or grain dishes, of course, constitute desserts.

Their place in the meal and the flavorings, sauces and garnishes that enhance them will determine what beverages accompany these dishes. Few of the presentations in this book are rarified enough to demand the finest wines as companions, and no wine tastes its best in the presence of very spicy or garlicky foods. Chinese fried rice or an Indian lentil purée, for example, is appropriately served with tea or beer, and beer is the drink to choose when offering hearty stews such as baked beans.

However, many bean or grain dishes can be happily paired with rustic, full-flavored red or white wines. A cassoulet, for example, would be complemented by sturdy Côtes-du-Rhône from France, or perhaps a Barbera from California. A young Spanish Rioja or a California Zinfandel makes a felicitous companion for paella. Risottos, often more subtle in flavor, can be mated with any number of wines; the decision should be based on the ingredients in the dish. A risotto that features seafood will be well served by a dry white Orvieto Secco from Italy, a Chenin Blanc from California or even a more delicate Sancerre from France. Vegetable risotto could be accompanied by Pinot Grigio, a white wine from Italy, and sausage risotto by Chianti Classico, a red wine from Italy, or perhaps Gamay Beaujolais, a red wine from California. In fact, there is as much scope for discovery with wines as there is with the preparations of the dishes; making your own judgments and experiments is part of the pleasure of learning about beans and grains.

The Multitudinous Pulses

Hundreds of varieties of beans, lentils and peas—collectively termed beans or pulses—are sold dried. Many of them are available in several different colors and, to add to the complexity, some have several names. A cross section of the pulses most commonly available in America is displayed below, together with a description of their flavor and texture after cooking. The samples are grouped by cat-egories—beans, lentils and peas—and ordered alphabetically within each cate-gory by the most common name. Alterna-tive names appear in parentheses.

All pulses are the seeds of leguminous plants—whose common feature is that the seeds are borne in pods. Each seed is made up of two sections encased in a strong skin. Physical shape is the most obvious distinction among beans, lentils and peas. Beans are generally oval or kidney shaped, lentils are flattened and disklike, and peas are round. The elon-gated black-eyed peas shown below are one exception, and are sometimes re-ferred to as black-eyed beans.

Some of these staples are sold in split form—usually without the skin, so that the two halves fall apart. Removal of the skin may produce a radical change in

Beans

Adzuki beans. Somewhat sweet flavor; soft texture.

Black beans. (Turtle beans.) Common beans with full, earthy flavor and mealy texture.

Cranberry beans. Common beans with mottled markings; nutlike flavor and mealy texture.

Fava beans. (Broad beans.) Assertive, almost bitter flavor and granular texture.

Navy beans. (Pea beans.) Common beans with mild flavor and mealy texture.

Pink beans. Common beans with meaty flavor and mealy texture.

Pinto beans. Mottled common beans with earthy, full-bodied flavor and mealy texture.

Red beans. Common beans with somewhat sweet flavor and mealy texture.

Lentils

Green lentils. Pronounced, somewhat pungent flavor; soft texture.

Pink lentils. Bland flavor; soft texture.

Split pink lentils. The split form of whole pink lentils.

Peas

Black-eyed peas. Succulent, earthy flavor and mealy texture.

appearance: For instance, when green mung beans are split and skinned, they yield a semispherical yellow bean known as golden gram.

Botanically, all lentils belong to one species, whereas beans and peas encompass representatives of several different species. But the majority of beans—the category that seems the most diverse—are, in fact, members of a single species, *phaseolus vulgaris,* or common bean.

Many of the beans, lentils and peas described below are available prepackaged—in boxes or plastic bags—at supermarkets or at health-food stores. Imported varieties, either prepackaged or sold by weight, may be obtained from markets that specialize in ethnic foods. When buying by weight, check that the beans, lentils or peas are of a uniform size and color, and free of mold. If prepackaging prevents examination, buy from a store with a quick turnover.

Although dried beans, lentils and peas have a long shelf life, you should avoid using any that are more than a year old: These staples darken and harden as they age, and require longer cooking. Eventually they dry out so much that they fall to pieces when cooked.

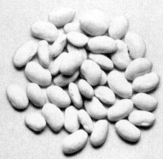

Great Northern beans. Common beans with mild flavor and mealy texture.

Kidney beans. Common beans available red or white; meaty flavor and mealy texture.

Lima beans. (Butter beans.) Available large or small; bland flavor and soft texture.

Mung beans. Available green *(above),* brown or black; somewhat sweet; soft texture.

Soybeans. Available yellow *(above),* green, brown, black or mottled; mild flavor; firm texture.

Flageolets. A French variety of common bean; delicate flavor and mealy texture.

Split black gram. (Urad dal.) Tangy flavor; soft texture.

Split golden gram. (Moong dal.) The split form of green mung beans.

Chick-peas. (Garbanzo beans.) Nutlike flavor; crunchy texture.

Field peas. Available green *(above)* or yellow; sweet flavor and soft texture.

Pigeon peas. Pungent flavor and mealy texture.

Split peas. The split form of garden or field peas.

Grains in Their Many Guises

All grain kernels are made up of four elements: a protein- or oil-rich embryo, or germ; an endosperm, composed mainly of starch and partially enclosing the germ; cellulose layers of bran, which sheathe the endosperm; and an inedible outer husk. After the husks are removed, the kernels may reach the market whole as groats—or, in the case of wheat, as berries—or processed to alter size, shape, and even flavor and texture. Displayed below are two dozen of the most readily

obtainable whole and processed forms, grouped alphabetically by grain. Each sample is identified by its most common name; where several names are widely used, alternatives are shown in parentheses. The flavor and texture of the grains after cooking are described in the first entry for each type.

The basic treatment for groats or berries is a friction process called polishing or pearling. This strips off the husk, as well as all or most of the bran, and also

removes the germ but leaves the endosperm whole. The mechanics of the process vary from grain to grain. The bran of cultivated rice can be taken off easily; with most varieties of barley, the bran is so tight that repeated pearlings with carborundum wheels are needed to remove it. The tough bran of corn may be removed by machine or by a long soaking in a caustic solution of lye or ground limestone and water, which attenuates the sweetness of the kernels to yield hominy.

Barley groats. Unpolished whole kernels of a barley hybrid; mild flavor; chewy texture.

Pearl barley. (Soup barley.) Whole kernels, polished four to six times; mild; tender.

Barley grits. Ground pearl barley; available in coarse, medium (above) and fine sizes.

Buckwheat groats. Whole unpolished kernels; mild and nutty flavor; soft texture.

Hominy grits. Ground yellow or white hominy; coarse, medium (above) or fine sizes.

Masa harina. Finely ground kernels of yellow or white hominy.

Millet. Unpolished whole kernels; cornlike flavor; firm texture.

Oat groats. Unpolished whole kernels; sweet, nutlike flavor; chewy texture.

Glutinous rice. Polished whole kernels of a short-grain variety; slightly sweet; extremely sticky.

Italian rice. (Arborio rice.) Polished whole kernels of a rounded short-grain variety; bland; soft texture.

Wild rice. Unpolished whole kernels; strong, nutty flavor; chewy texture.

Cracked rye. Cracked unpolished kernels; slightly sour taste; soft texture.

Regardless of the process used, polished kernels cook more quickly than groats and have a smoother texture. As alternative ways to speed their cooking, groats or polished kernels are cracked or sliced into smaller pieces, and wheat berries are steamed until soft, dried again and then cracked to make bulgur.

Cracked grain may be ground coarse to form grits, or ground fine to form meals and flours. In another variation, whole, cracked or sliced groats are heated and pressed flat to form flaked or rolled grain.

Groats, because they retain the oily germ, are more susceptible to rancidity than polished kernels. And in cracking, pressing or grinding, the groats' germs are broken open, thus allowing the oil to escape and increasing the product's susceptibility. For this reason, both whole and processed forms should be bought in small quantities and stored, tightly covered, in a cool place, preferably the refrigerator. Even so, they have a storage life of no more than four or five months.

Polished kernels—whether whole or cracked, ground or pressed—last longer. In tightly sealed containers, they may be kept in a cool, dry place for a year.

Of the grain products shown below, most are widely available either at supermarkets or in health-food stores. Look for masa harina where Latin American foods are sold; glutinous rice is obtainable at Asian food markets and Italian rice at Italian food or gourmet shops.

Toasted buckwheat groats. (Whole kasha.) Oven-toasted to intensify flavor; soft texture.

Toasted buckwheat grits. (Kasha.) Ground groats; coarse, medium *(above)* or fine.

Cornmeal. Ground yellow or white kernels, usually polished; sweet.

Cracked hominy. Cracked white or yellow kernels, soaked to remove bran; mildly sweet; firm.

Rolled oats. (Oat flakes.) Steamed and flattened, whole or sliced oat groats.

Oatmeal. Ground oat groats; available in coarse, or "pinhead," medium, and fine *(above)* sizes.

Brown rice. Unpolished long *(above)*, medium or short kernels; nutty flavor; firm texture.

White rice. Polished firm-textured long *(above)* or softer medium or short kernels; bland flavor.

Wheat berries. Unpolished whole kernels; robust flavor; very chewy texture.

Cracked wheat. Cracked wheat berries; available in coarse, medium *(above)* and fine sizes.

Bulgur. Steamed, dried and cracked wheat berries; mild nutty flavor; soft texture.

Semolina. Ground polished kernels of durum wheat—an especially hard grain.

1
Cooking with Water
Variations on a Simple Theme

Pearly short-grain rice, destined to be the base for a Milanese risotto (page 26), is poured into a pan where onion and beef marrow already stew in butter. To achieve the risotto's characteristic sticky consistency, hot stock will be added a little at a time, and the grains will be regularly stirred.

Whether whole or split, cracked or ground, dried beans and grains must absorb water to become plump, tender and succulent. Boiling, simmering, steaming and steeping provide a spectrum of methods suitable for rehydrating these staples. The water used may be plain, or salted, or enhanced with herbs, spices and aromatic vegetables or meats; it may even be transformed ahead of time into a rich stock (page 166). And the options for serving the rehydrated beans and grains are at least as varied. Since grains and beans harmonize with almost any vegetable, fruit, fish, shellfish or meat, the cook can freely improvise accompaniments.

No matter how they are to be presented, most dried beans and peas require lengthy cooking—often preceded by soaking—to swell them fully (their size increases about two and one half times). Lentils cook more quickly because they are small and have relatively thin skins, but like whole beans and peas, they demand gentle simmering and a minimum of stirring to keep them intact. Even careful treatment may not prevent split beans, lentils and peas from disintegrating; it is best to make a virtue of necessity and serve them as purées. For textural variety, any cooked bean or pea can, of course, be crushed to a coarse purée and then sieved to melting smoothness.

Grains lend themselves to more cooking methods and inspire more culinary ingenuity than beans or peas. Depending on the amount of water used and the character of the grain, whole kernels and some cracked ones may be boiled or simmered until they are fluffy and dry, or soft and moist, or sticky enough to hold with chopsticks (pages 22-23). Two classic rice dishes dramatize the possibilities: For the tender, separate kernels of a pilaf, long-grain rice is heated in fat, then simmered undisturbed in precisely the amount of water it can absorb without becoming sticky (pages 24-25). For a risotto, the stickiness of short-grain rice is exploited by boiling and stirring—and the creaminess of the end result is intensified with cheese and butter.

Steaming and steeping further extend the cook's options. If cooked above water in a steamer, cracked or ground grains soften but remain fluffy; if boiling water is poured over them and they are left to soak, grains swell but remain chewy. By any method, however, grains expand more than beans do—often to three, even four, times their dried volume.

The Fundamentals of Bean and Grain Cookery

Because beans, lentils, peas and grains vary so widely in size, texture and form, the methods by which they are cooked in water can range from vigorous boiling to gentle steaming. The box below details the methods that suit each type—and the advance treatments some require.

Lentils and split beans or peas need only be cleaned before they are cooked (pages 16-17). However, whole beans and peas must be soaked in water after they are cleaned: Their skins are impermeable so water can only enter through the

hilum, the point at which each bean or pea was attached to its pod. Therefore, rehydrating the flesh within the skin is a slow process, and soaking is insurance that beans and peas will cook evenly.

Generally, beans or peas are soaked in cold water for at least eight hours and up to 12 hours or overnight. A more expeditious method—called quick soaking—is to bring the water and beans or peas to a boil, then boil them for two minutes, cover them and set them aside to soak for an hour. In either case, the water is likely to develop a somewhat sour flavor and is best discarded.

To destroy toxins (box, opposite) and to preserve the texture of the beans, lentils and peas, all of them should be brought slowly to a boil and, in some cases, boiled for 10 minutes. They are then covered and simmered until tender. Even split lentils and peas, which cook naturally to a purée, will have a smoother consistency if simmered slowly. The cooking times shown below are approximations; always test for doneness by squeezing a whole bean or pea or tasting a dollop of purée.

Unlike beans, only a few grains need

A Guide to Times and Treatments

Beans

Beans, adzuki. Soak in cold water for eight hours or overnight; or quick-soak for one hour. Boil for 10 minutes in three parts water to one part beans and simmer, covered, for one and a half to two hours.

Beans, black. Soak in cold water for eight hours or overnight; or quick-soak for one hour. Boil for 10 minutes in three parts water to one part beans and simmer, covered, for one to one and a half hours.

Beans, cranberry. Soak in cold water for eight hours or overnight; or quick-soak for one hour. Boil for 10 minutes in three parts water to one part beans and simmer, covered, for one and a half to two hours.

Beans, fava. Soak in cold water for eight hours or overnight; or quick-soak for one hour. Peel off the bitter brown skins. Boil for 10 minutes in three parts water to one part beans and simmer, covered, for two to two and a half hours.

Beans, Great Northern. Soak in cold water for eight hours or overnight; or quick-soak for one hour. Boil for 10 minutes in three parts water to one part beans and simmer, covered, for one and a half to two hours.

Beans, kidney. Soak in cold water for eight hours or overnight; or quick-soak for one hour. Boil for 10 minutes in three parts water to one part beans and simmer, covered, for one and a half to two hours.

Beans, lima. Soak in cold water for eight hours or overnight; or quick-soak for one hour. Boil for 10 minutes in three parts water to one part beans and simmer,

covered, for one to one and a half hours.

Beans, mung. Soak in cold water for one to four hours. Drain. Bring to a boil in three parts water to one part beans and simmer, covered, for 30 minutes.

Beans, navy. Soak in cold water for eight hours or overnight; or quick-soak for one hour. Boil for 10 minutes in three parts water to one part beans and simmer, covered, for one and a half to two hours.

Beans, pink. Soak in cold water for eight hours or overnight; or quick-soak for one hour. Boil for 10 minutes in three parts water to one part beans and simmer, covered, for one and a half to two hours.

Beans, pinto. Soak in cold water for eight hours or overnight; or quick-soak for one hour. Boil for 10 minutes in three parts water to one part beans and simmer, covered, for one and a half to two hours.

Beans, red. Soak in cold water for eight hours or overnight; or quick-soak for one hour. Boil for 10 minutes in three parts water to one part beans and simmer, covered, for one and a half to two hours.

Flageolets. Soak in cold water for eight hours or overnight; or quick-soak for one hour. Boil for 10 minutes in three parts water to one part beans and simmer, covered, for one and a half to two hours.

Gram, split. Bring to a boil in four parts water to one part beans and simmer, covered, for 20 to 30 minutes for a purée.

Soybeans. Soak in cold water for eight hours or overnight; or quick-soak for one hour. Boil for 10 minutes in three parts water

to one part beans and simmer, covered, for one and a half to two hours.

Lentils

Lentils, split. Bring to a boil in three parts water to one part lentils and simmer, covered, for 20 to 25 minutes for a purée.

Lentils, whole. Bring to a boil in three parts water to one part lentils and simmer, covered, for 20 to 25 minutes for whole lentils, or 30 to 35 minutes for a purée.

Peas

Chick-peas. Soak in cold water for eight hours or overnight; or quick-soak for one hour. Bring to a boil in three parts water to one part peas and simmer, covered, for one and a half to two hours.

Peas, black-eyed. Soak in cold water for eight hours or overnight; or quick-soak for one hour. Bring to a boil in three parts water to one part peas and simmer, covered, for about one hour.

Peas, field. Soak in cold water for eight hours or overnight; or quick-soak for one hour. Bring to a boil in three parts water to one part peas and simmer, covered, for about one hour.

Peas, pigeon. Soak in cold water for eight hours or overnight; or quick-soak for one hour. Bring to a boil in three parts water to one part peas and simmer, covered, for one and a half to two hours.

Peas, split. Bring to a boil in four parts water to one part peas and simmer, covered, for about 30 minutes for whole split peas, or 40 to 45 minutes for a purée.

to be soaked. Glutinous rice requires lengthy soaking to ensure that it will cook evenly; cracked hominy is soaked to speed the tenderizing of its tough kernels. Rinsing under cold running water cleans the kernels of wild rice and removes surface starch from white rice so that the kernels do not become too sticky when cooked. However, nutritionists argue against applying the tactic to white rice because soluble iron and B vitamins that are sprayed onto the surfaces of most American-grown varieties will be lost.

Although all beans are cooked by the same basic method, unprocessed grains lend themselves to various strategies, each of which produces slightly different results: boiling in as much water as a pot will hold; simmering in measured water; steaming above water; and steeping by soaking the kernels in water that has been brought to a boil. Grains that have been processed by parboiling should be cooked only according to the manufacturer's package directions. With any method, use timing instructions as guidelines, but taste grains to test them for doneness before serving them.

Grains

Barley, pearl. Boil, uncovered, in unlimited water or simmer, covered, in measured water—two parts water to one part barley—for 45 minutes. Or steep in one part boiling water to two parts barley for 30 minutes.

Barley grits. Simmer, covered, in measured water—two parts water to one part grits—for 15 minutes. For porridge, simmer uncovered in four parts water to one part grits for 25 minutes. Or steep in one part boiling water to two parts grits for 30 minutes. Or steam for one hour.

Barley groats. Boil, uncovered, in unlimited water or simmer, covered, in measured water—two parts water to one part groats—for one hour.

Buckwheat grits, roasted. Simmer, covered, in measured water—two and one half parts water to one part grits—for 10 to 12 minutes. For porridge, simmer uncovered in five parts water to one part grits for 25 minutes.

Buckwheat groats, plain or roasted. Simmer, covered, in measured water—two parts water to one part groats—for 15 minutes.

Bulgur. Simmer, covered, in measured water—two parts water to one part wheat—for 30 to 40 minutes. For porridge, simmer uncovered in four parts water to one part wheat for 30 minutes. Or steep in one part boiling water to two parts wheat for 30 minutes. Or steam for one hour.

Cornmeal. For porridge, simmer uncovered in four parts water to one part meal for 25 minutes.

Hominy, cracked. Soak for eight hours or overnight. Boil, uncovered, in unlimited water or simmer, covered, in measured water—three parts water to one part hominy—for two and a half to three hours.

Hominy grits. For porridge, simmer uncovered in four and a half parts water to one part grits for 25 minutes.

Millet. Boil, uncovered, in unlimited water or simmer, covered, in measured water—two parts water to one part millet—for 30 to 35 minutes.

Oat groats. Boil, uncovered, in unlimited water or simmer, covered, in measured water—two parts water to one part groats—for 45 minutes to one hour.

Oatmeal. For porridge, simmer uncovered in three parts water to one part meal for 25 minutes.

Oats, rolled. For porridge, simmer uncovered in two parts water to one part oats for 10 minutes.

Rice, brown. Boil, uncovered, in unlimited water or simmer, covered, in measured water—two parts water to one part rice—for 45 minutes to one hour.

Rice, glutinous. Soak for 12 hours or overnight. Drain. Simmer, covered, in measured water—one part water to one part rice—for about 12 minutes.

Rice, Italian. Rinse (optional). Boil, uncovered, in unlimited water or simmer, covered, in measured water—one and a half parts water to one part rice—for 15 minutes.

Rice, long-grain white. Rinse (optional). Boil, uncovered, in unlimited water for 12 to 15 minutes. Or simmer, covered, in measured water for 15 minutes—using one and three quarters parts water to one part rice for a moderately soft cooked product or two parts water to one part rice for a soft product.

Rice, medium- or short-grain white. Rinse (optional). Boil, uncovered, in unlimited water for 12 to 15 minutes. Or simmer, covered, in measured water for 15 minutes—using one and a half parts water to one part rice.

Rice, wild. Rinse. Boil, uncovered, in unlimited water or simmer, covered, in measured water—three parts water to one part rice—for one hour.

Rye, cracked. Simmer, covered, in measured water—two parts water to one part rye—for about 15 minutes. For porridge, simmer uncovered in four parts water to one part rye for 25 minutes. Or steam for one hour.

Semolina. For porridge, simmer uncovered in three and one half parts water to one part grain for 25 minutes.

Wheat berries. Boil, uncovered, in unlimited water or simmer, covered, in measured water—two parts water to one part wheat—for one hour.

Wheat, cracked. Simmer, covered, in measured water—two parts water to one part wheat—for 30 to 40 minutes. For porridge, simmer uncovered in four parts water to one part wheat for 25 to 30 minutes. Or steep in one part boiling water to two parts wheat for 30 minutes. Or steam for one hour.

Basic Handling for Beans, Lentils and Peas

Gentle simmering in water turns dry, hard beans, lentils or peas into plump and succulent nuggets. During cooking, the bland beans may be enriched by adding meat and aromatics to the water, as shown here, or by replacing the water with stock (page 25). Afterward, moisten the beans by dressing them with butter, olive oil or the tangy vinaigrette used opposite, at bottom. If they are to accompany roast meats, they can be dressed with degreased pan liquids.

To augment the dressing, beans can be garnished in a multitude of ways. Among the options are chopped or sieved hard-boiled eggs, chopped fresh herbs such as basil or chives, sautéed onion rings or—as shown below—chopped cooked meat and chunks of peeled and seeded tomatoes that have been sautéed in butter.

However you plan to serve them, all beans need to be cleaned before cooking can begin. First spread them on a tray or large plate and pick them over. Beans sold in bulk often contain bits of foreign

matter. Even prepackaged beans may include discolored or broken specimens.

Next immerse the beans briefly in water. Beans that are harvested when immature may mold, or they may shrink within their skins, thus creating an air pocket inside that can harbor dirt. Such defects are not visible, but they make the beans buoyant enough to float to the top of the water, where they may be removed easily, along with any loose skins. To complete the cleaning process, rinse the beans in a colander.

At this stage, whole beans and peas should be soaked, but lentils and split beans or peas may be immediately covered with water, brought to a boil and simmered until tender. As they approach the boiling point, all beans, lentils or peas will release some starch, protein and minerals in the form of a grayish scum. Although this scum will eventually disperse through the cooking water, most cooks remove it lest it leave a gritty film on the cooked beans.

An Addition of Aromatics

1 **Rinsing.** Pick over the dried beans, lentils or peas—here, black-eyed peas—and discard any discolored specimens or bits of grit. Place the beans in a bowl, cover them with cold water and, after a minute or so, remove any skins or whole beans that float to the surface. Rinse the beans in a colander set under cold running water.

5 **Adding aromatics.** To flavor the cooking liquid, add aromatic vegetables—in this case, carrots, an onion studded with whole cloves and an unpeeled garlic bulb—pushing them down among the beans, lentils or peas. Make a bouquet garni by tying together fresh sprigs of thyme and parsley, a bay leaf and a celery rib. Add the bouquet garni to the pot, then reduce the heat so that the beans simmer gently.

6 **Simmering and serving.** Add boiling water if more liquid is necessary, but do not stir the beans, lentils or peas. Cover, and simmer them for about one hour, or until one feels tender when squeezed (inset). Discard the aromatics; drain, skin, bone and chop the ham. Drain the beans, toss them with butter and serve garnished with the ham and, if you like, sautéed tomatoes (above).

2 **Soaking.** Using a bowl for slow soaking or a pot for quick soaking *(pages 14-15)*, immerse the beans in cold water—allowing 3 cups [¾ liter] of water to 1 cup [¼ liter] of beans. Soak them for the time listed in the guide. Drain the beans and discard the water.

3 **Salting.** Put the drained beans, lentils or peas in a heavy pot and add 3 cups of water for each 1 cup of beans. The water should cover the beans by at least 1 inch [2½ cm.]. Add ½ to 1 teaspoon [2 to 5 ml.] of salt for each cup of beans or, as shown, add a ham hock or a piece of another salt-cured meat such as slab bacon or salt pork.

4 **Skimming.** Bring the beans, lentils or peas slowly to a boil over medium heat. Using a ladle or skimmer, lift off the scum that will rise to the surface. Repeat this process until the scum stops rising. Boil beans—but not lentils or peas—for 10 minutes.

Vinaigrette to Finish Lentils

1 **Dressing.** Clean green and pink lentils *(Step 1, opposite)*. Using separate pots, immerse them in water, add salt, bring to a boil, skim, reduce the heat and simmer for 20 to 25 minutes, until tender. Drain the lentils and place them in separate bowls. While they are still hot and absorbent, toss them with a mustard-flavored vinaigrette, made by combining four parts of oil with one part of vinegar.

2 **Marinating.** Turn the lentils about in the vinaigrette until they are evenly coated. Set the bowls aside to allow the lentils to cool and marinate for at least half an hour. Then ladle the lentils onto a serving plate, arranging them in separate strips.

3 **Garnishing.** Hard-boil and shell two eggs. Separate the egg whites from the yolks and press the yolks through a fine-meshed nylon or stainless-steel sieve. Chop the whites into small bits. Sprinkle the sieved yolks over the lentils and decorate the edges of the serving plate with the whites.

A Choice of Purées

The soft flesh of cooked beans, lentils and peas yields purées quite diverse in character: The purées may be flavored and enriched in any of a number of ways, and the method of preparing them can be altered to produce textures ranging from rough and chunky to creamily smooth.

The beans can be cooked in water or in meat stock *(page 25);* for complexity of flavor, the cooking liquid may be augmented with aromatics such as garlic, ginger, carrots or onions, and with such herbs as bay leaves, thyme, tarragon and sage. The purées of India often include spices such as turmeric, coriander, ginger or cumin.

Once softened by soaking or cooking, the beans, lentils or peas are puréed in any of three ways, the choice depending both on what you are puréeing and on the texture you desire. Lentils and split beans and peas are so soft after cooking that they can simply be whisked to a coarse purée *(right).* Whole beans or peas such as Great Northern beans or chick-

peas can be pounded to a purée, but the texture will be extremely rough; it is easier to purée them with the aid of a food processor or food mill.

Most of these purées contain the skins of the beans or peas. In some cases, skins are desirable: They lend a rough purée its character and contribute their colors to smooth purées such as those shown on pages 64-65. The palest, smoothest purées, however, contain no skins. To give any purée this delicate texture, force it through a fine-meshed sieve, as shown opposite, below.

Enrichments—typically oil or butter —and additional flavorings can be added to the beans, lentils or peas after puréeing is complete. The Indian lentil purée shown here, for instance, is brightened with a little lemon juice and further enhanced by oil—in this case, vegetable oil—with garlic and cumin seeds. The smooth chick-pea purée shown opposite is enriched with butter that has been beaten with fresh tarragon.

A Coarse Whisked Texture

1 **Preparing lentils.** Pick any grit or stones from lentils—skinned pink lentils are used for this demonstration. Wash the lentils, drain them *(pages 16-17)* and place them in a saucepan. Add peeled, finely chopped fresh ginger; stir in ground turmeric.

4 **Preparing seasonings.** Heat oil in a skillet; stir in cumin seeds and fry them over high heat for about 10 seconds, until fragrant. Remove the skillet from the heat and stir in crushed garlic cloves and cayenne pepper.

5 **Serving the purée.** Stir the garlic and pepper in the oil for a moment—just until the garlic begins to color. Do not let the garlic turn dark brown. Immediately pour the perfumed oil over the lentil purée. Just before serving, stir the oil into the lentils.

2 **Cooking the lentils.** Add liquid—water is used here—to the pan, allowing 3 to 3½ cups [750 to 875 ml.] of liquid for each cup [¼ liter] of lentils. Set the pan over medium heat and bring the liquid to a boil, stirring frequently to keep the lentils from sticking to the pan.

3 **Puréeing the lentils.** As soon as the liquid reaches a boil, reduce the heat to low, partly cover the pan and simmer the lentils for 20 to 25 minutes, until they have softened and begun to disintegrate. Remove the pan from the heat and begin to beat the lentils with a whisk *(above, left)*. Continue to beat while you squeeze in fresh lemon juice *(right)* and stir in salt to taste. Stop beating when the lentils have been reduced to a coarse purée.

Sieving for Smoothness

1 **Grinding the chick-peas.** Soak chick-peas overnight, drain them and cook them until they are tender *(pages 16-17)*; in this case, the chick-peas were cooked in meat stock. Drain the chick-peas and grind them through a food mill fitted with a medium disk to make a coarse purée. Alternatively, purée the drained peas in a food processor.

2 **Sieving the chick-peas.** Force the chick-pea purée through a fine-meshed sieve—a drum sieve is used here. Transfer the fine purée thus formed to a pan set over medium heat and stir in cooking liquid to dilute the purée to the consistency of a thick sauce.

3 **Serving.** Heat the purée for about five minutes, stirring frequently to prevent sticking. When the purée is steaming hot, transfer it to a heated serving bowl and stir in softened butter flavored with chopped fresh herbs—tarragon, in this instance. Serve the purée as soon as the butter melts.

Easy Steps to Flawless Bean Curd

Puréed soybeans provide a unique dividend: The purée—made from beans that are soaked but not cooked—will yield a liquid known as soybean milk. Properly handled, soybean milk solidifies into curds, as cow's milk does in cheese making. When these curds are compressed, they form the silky-textured, subtly flavored cakes featured in many Japanese and Chinese dishes and known variously as bean curd, *tofu* and *dou fu*. The cakes can be bought ready-made at many supermarkets, of course, but it is not hard to prepare them at home *(recipe, page 165)*. And homemade cakes are incomparably fresh in flavor and fine in texture.

The ingredients for making the cakes are soybeans—the only readily available beans high enough in the protein necessary for curd formation and low enough in starch to prevent the milk from becoming gummy—and a solidifier. The solidifier often used by Asian cooks is *nigari;* sold at Asian markets and health-food stores, *nigari* is a granular substance high in magnesium chloride. Alternatively, you can use magnesium sulfate, a salt found at pharmacies. Either solidifier causes the protein molecules in the soybean milk to link together, producing curds. For bean curd with a tangy taste, use lemon or lime juice or vinegar, whose acids also cause curdling.

No special equipment is needed to extract soybean milk and shape its curds. However, two devices shown here—and obtainable in health-food stores—will simplify the work. The first is a muslin sack to hold the purée while you press out its milk; lacking a sack, you can wrap the purée in cheesecloth. The second device is a perforated wood box with a removable base and lid, used for pressing the curds into cakes; a bread pan with holes punched or drilled in the bottom and sides can serve instead.

Once it has reached the desired consistency *(Step 6)*, the bean curd may be marinated and served cold: Its mild flavor is a perfect foil for spicy elements. Or it may be stir fried with meats, vegetables and seasonings *(pages 50-51)*.

1 **Heating bean purée.** Soak soybeans overnight in cold water *(pages 14-15)*; do not use the quick-soaking method. Drain the beans and purée them with a little fresh water in a food processor. Put the purée in a heavy pot containing more water, bring to a boil over moderately high heat and cook for 10 minutes, stirring occasionally.

4 **Forming a cake.** Line a perforated mold—a wood pressing box is used here—with dampened cheesecloth, leaving the ends of the cloth long. Set the box in a baking dish to catch pressed-out liquid. Using a perforated spoon, ladle the curds into the box.

5 **Weighting the cake.** Fold the ends of the cheesecloth over the top of the soybean curds, covering them completely; set a board or the flat box lid on the curds. Place a 4-pound [2-kg.] weight—in this case, a bean-filled jar—on the box lid. For fairly soft bean curd suitable for marinating *(Step 6)*, let the bean curd rest at room temperature for 45 minutes. For firm bean curd suitable for frying *(pages 50-51)*, let the bean curd rest for two to three hours.

2 **Draining the milk.** Set a colander in a large bowl; line the colander with moistened cheesecloth or, as here, with a moistened muslin sack. Ladle in the hot purée and let it cool until you can touch it. Close the cloth or sack around the purée, and squeeze and press to force the milk into the bowl. Throw away the pulp.

3 **Curdling the milk.** Pour the soybean milk into a large pot; reheat the milk until it boils and remove it from the heat. If necessary, dissolve the solidifying agent—here, magnesium chloride—in water. Stir the soybean milk while you add a third of the solidifier *(above, left)*. Let the mixture rest, covered, for three minutes. Add a second portion of solidifier and stir the top of the liquid; curds will begin to form *(center)*. Let the mixture rest for three more minutes. Stir in the last of the solidifier *(right)*; the curds will separate from the yellow whey.

6 **Marinating.** The weighted cake may be unmolded and used immediately, or unmolded, placed in a water-filled container and refrigerated for up to one week if the water is changed daily. To unmold the cake, lift off the weight and lid. Remove the sides of the box—invert any other kind of mold—and unwrap the cloth. To marinate the cake, cut it into broad slices and place them in a marinade—a mixture of vinegar, chopped dried hot chilies, soy sauce, sesame-seed oil and Szechwan pepper is used here.

7 **Serving the bean curd.** Let the bean curd marinate at room temperature for about half an hour, turning the slices once. Meanwhile, trim fresh spinach leaves, blanch them in boiling water for 30 seconds, refresh them under cold running water and pat them dry with paper towels. Arrange the leaves on a serving dish. Cut each marinated bean-curd slice crosswise at ⅓-inch [1-cm.] intervals and, keeping the cut-up slices intact, place them on the leaves. Sprinkle a little of the marinade over the bean curd.

Three Ways to Tenderize Grain

Almost 70 per cent of a grain kernel is starch, and when the kernel absorbs hot water and its starches swell, it puffs up and is transformed into a tender, translucent morsel. The process can be accomplished in three ways, demonstrated here with long-grain white rice. The method you choose will depend on the character of the grain used and the effect desired.

The simplest strategy, and the one yielding the fluffiest, driest grains, is to boil them in unlimited water. All whole grains, from rice to millet, can be cooked this way. To prevent lumping, bring the water to a boil before adding the kernels and stir them well immediately. But do not stir again, lest the cooking kernels release particles of surface starch and become sticky.

Since more water is used than can be absorbed, the kernels must be drained. Then they should be dried and allowed to finish softening by assimilating any water still on their surfaces—a step that can be done in the oven, as shown here, or in a heavy pot set over very low heat.

For somewhat moister kernels, grains may be simmered in measured water. The water is absorbed by the kernels as they cook, leaving them ready to serve after they have rested and been fluffed.

Simmering in measured water is an appropriate method for all cracked as well as whole grains—with one notable variation for buckwheat groats. Because these groats have especially tender hulls, they should be coated with beaten egg—allow one egg for each cup [¼ liter] of groats—and cooked for a minute or so to set the egg before water is added.

A third strategy, for white rice only, is the Chinese method shown at far right. Designed to yield kernels that cling to chopsticks, this method also uses measured water. However, the rice is stirred as soon as it has absorbed most of the water; this releases starch particles so that the kernels develop a soft—but not gummy—texture in the final cooking.

Whatever strategy is adopted, most grains swell to about three times their original volume. White and brown rice, high in gelatinizing substances known as amylose or amylopectin, will expand three to four times.

Boiling in Unlimited Water

1 **Immersing the grain.** Fill a large saucepan nearly full of salted water and bring it to a boil. Adjusting the heat to keep the water at a boil, gradually sprinkle in the grain—in this case, long-grain white rice. Stir once with a fork or spoon to separate the kernels. To keep the water from foaming over, add a spoonful of flavorless oil to the pan.

2 **Draining the grain.** Boil the grain uncovered for 12 minutes (pages 14-15), or until it has expanded to about three times its original volume and is softened but still slightly resilient. To test the grain, remove a few kernels and bite them or pinch them. Strain the grain in a strainer or colander.

Simmering in Measured Water

1 **Adding water.** Place a measured amount of grain in a saucepan and add a measured amount of cold water. In this instance, the ratio is 1 cup [¼ liter] of long-grain white rice to 1¾ cups [425 ml.] of water. Sprinkle in a little salt and stir with a fork or spoon.

2 **Simmering.** Bring the water to a boil over medium heat, cover and reduce the heat to low so that the water simmers slowly. Without stirring, simmer the grain for about 15 minutes, or until the kernels have swollen to about three times their original size and all of the water has been absorbed.

3 Drying and serving the grain. Tip the grain into a shallow baking dish and spread it out evenly. Cover the grain and put it in a preheated 300° F. [150° C.] oven for 10 to 15 minutes, until it is completely tender and dry. Add a few pieces of butter and toss the grain, using forks to avoid crushing the kernels. Serve at once.

3 **Resting and serving.** Take the saucepan off the heat, wrap the lid with a cloth towel and set the lid back on the pan *(inset)*. The towel will absorb the moisture that rises from the grain. Let the grain rest in the pan for five to 10 minutes, then fluff it with two forks *(above)* and serve it immediately.

A Chinese Tactic

1 **Stirring the rice.** Combine the rice with twice its volume of cold water. Bring to a boil over high heat, cover and boil for three minutes more, or until the water level matches the level of the rice and small craters appear in the surface. Stir the rice thoroughly, making sure that no kernels stick to the pan.

2 **Fluffing the rice.** Cover the pan and cook the rice over moderately low heat for 15 minutes, or until the grains are very soft and have swollen to about three times their original volume. Fluff the rice with forks or chopsticks and serve immediately.

Pilaf: Preliminary Sautéing to Enrich the Grain

One of the easiest ways to obtain soft but separate kernels is to heat grain in a little oil or fat, then simmer it, tightly covered, in a measured amount of liquid. The fat helps to keep the kernels from sticking together, and gentle simmering cooks them to tenderness. Grain dishes prepared by this technique are known to Western cooks as pilafs, and in Western cookery, the term is often reserved for white rice. However, almost any whole or cracked grain—barley, buckwheat, bulgur, millet, brown rice, wild rice, cracked wheat—can be cooked in this manner.

Care must be exercised in choosing the cooking medium for the initial sautéing of the grain. Vegetable oils, which can safely be heated to high temperatures, are a better choice for sautéing than plain butter, which will scorch before the grain can absorb it. However, if you want a buttery taste, you can use butter melted in vegetable oil in a proportion of two to one. Or you can clarify the butter beforehand to remove its easily burned milk solids. To do so, melt the butter over low heat, then let it stand off the heat until the foam rises to the top and the white milk solids settle. Skim off the foam and decant the clear yellow butter fat, leaving the solids behind.

Once coated with fat, the grain is simmered in water or, for flavor, in meat stock (box, opposite). In either case, the ratio of liquid to grain should be slightly less than two to one: For 1 cup [¼ liter] of grain, allow 1¾ cups [425 ml.] of liquid.

These proportions may be adjusted according to the other ingredients cooked with the grain. Many pilafs, for instance, are flavored with spices and vegetables, added during sautéing to bring out their flavors. Other fresh vegetables can be added during the simmering period—at intervals suited to the amount of cooking they need. All fresh vegetables will release liquid; if you use them, start with a little less liquid than usual, adding more if the grain begins to look dry.

Garnishes for pilafs appear in bewildering variety. In the top demonstration, tomatoes contribute color and sweetness to an elemental pilaf. The more complex millet pilaf in the bottom demonstration is enhanced by fried onions and ginger (recipe, page 154).

A Finishing Touch of Tomato

1 Sautéing rice. In a saucepan set over medium heat, warm oil—here, vegetable oil—until a drop of water thrown into the pan spatters. Pour in long-grain rice. Stirring gently, sauté the rice for two to three minutes, until most of the grains turn a milky color.

2 Adding liquid. Pour a measured amount of boiling water—heated so that it will not slow the cooking—into the pan. Salt the rice and stir it once to loosen any grains that may cling to the pan. Return the water to a boil, reduce the heat so that the water just simmers, and put the lid on the pan.

An Intrinsic Enhancement of Vegetables

1 Sautéing flavorings. Place a pepper under a preheated broiler for about 10 minutes, turning it frequently to char and loosen its skin evenly. Peel, halve, seed, derib and chop the pepper. Sauté it with chopped onion, sliced garlic, cumin seeds and black mustard seeds in butter and oil for about 10 minutes, until the onion is translucent.

2 Sautéing millet. Cut unpeeled potatoes into cubes. Remove the core and leaves from a cauliflower, then trim it into small florets. Add the potatoes and cauliflower to the sauté pan and cook them for 10 minutes over medium heat, stirring frequently. Add millet and sauté it for about three minutes, stirring to coat it with the butter and oil.

A Versatile Meat Stock

To make a hearty stock, suitable for cooking any kind of grain, combine aromatic vegetables—onions, carrots, leeks and celery, for example—with herbs and inexpensive cuts of meat. Here, beef provides rich flavor, while chicken and veal add a slight sweetness *(recipe, page 166)*. Cook the meat and vegetables in water until their essences are drawn out, then strain the liquid.

3 **Fluffing the rice.** Simmer the rice undisturbed for about 15 minutes, or until all of the water has been absorbed. Remove the pan from the heat. Take off the lid, wrap it with a towel, replace the lid, and allow the rice to rest for 10 minutes. Add butter and fluff the rice with forks. It may be served at this point, or a garnish may be added *(Step 4, right)*.

4 **Adding tomatoes.** Core tomatoes and blanch them in boiling water for 20 seconds to loosen their skins. Transfer the tomatoes to a bowl of cold water to stop the cooking. Then peel them, halve them and dig out the seeds. Chop the tomatoes into coarse chunks and stew them in a little butter in a pan set over medium heat. Add salt and pepper, then gently mix the tomatoes into the rice pilaf.

1 **Adding vegetables.** Put the meat on a rack in a pot and cover with cold water. Bring it slowly to a boil and remove the scum. Add vegetables and herbs; skim. Put the lid on the pot, slightly ajar.

3 **Simmering the millet.** Pour hot stock into the pan and bring it to a boil. Reduce the heat to low so that the liquid just simmers, cover the pan and cook the millet for about 35 minutes. Check periodically to make sure that the millet does not dry up; if it appears dry, ladle in a little hot stock. When the millet is tender and fluffy, remove the sauté pan from the heat.

4 **Garnishing.** Peel and slice onions and separate them into rings. Peel ginger and sliver it. Sauté the onions in clarified butter over medium heat for five to 10 minutes, until the rings brown. Add the ginger and sauté for about two minutes. Drain the onion mixture, reserving the butter; pour it over the millet pilaf. Fluff the millet with forks and garnish it with the onions and ginger.

2 **Straining.** Adjust the heat so that the liquid barely simmers. Cook the stock for four to five hours. Strain it through a colander lined with cheesecloth. Refrigerate the stock until the fat rises and solidifies on its surface; scoop off the fat.

Risotto: Grains Suspended in a Creamy Liquid

The risotto of Italy might be described as a form of pilaf *(pages 24-25):* To make it, rice is sautéed in fat or oil, then cooked in liquid. However, a risotto is quite different in character from a pilaf: Instead of containing fluffy, separate grains, the dish consists of a creamily homogeneous, pourable mass. The difference is caused both by the grain used and by the method of simmering it.

A perfect risotto requires Italian rice, a short-grain variety that is high in amylopectin—a gelatinizing component of starch—and therefore especially clinging. The delectably smooth nature of the finished dish also depends on adding liquid—stock is invariably used for flavor—to the sautéed rice in small increments, stirring all the while. The liquid is added in this manner so that the kernels are able to slowly absorb the maximum amount of liquid without becoming soft and mushy; as a rule, the rice is cooked with about three times its own volume of stock. Stirring also causes the

surface starch to dissolve, producing a saucelike medium for the grains.

Like pilaf, risotto can be enhanced in any number of ways *(recipes, pages 139-140).* Onions usually are sautéed with the rice for flavor; the choice of additional garnishes is almost unlimited. The Milanese risotto demonstrated below, for instance, includes beef marrow, wine and saffron. The Venetian risotto shown opposite, below, is enhanced by ham and peas, which give it its Italian name—*risi et bisi,* or "rice and peas." Other risottos contain parboiled vegetables, sautéed meats, or shellfish.

Whatever their contents, almost all risottos receive a final garnish that adds to their creaminess—such as chunks of softened butter and generous quantities of freshly grated Parmesan cheese. All risottos should be served immediately: Their creamy quality disappears as they cool. Any leftover risotto may be used to form the croquettes that are demonstrated on pages 56-57.

Increments of Stock

1 **Sautéing the rice.** Melt butter and vegetable oil in a large saucepan set over medium heat. Add finely chopped onion and sauté it for about 10 minutes, until it is translucent. Pour in short-grain rice and heat for two to three minutes, stirring to ensure that all of the grains are coated with fat.

A Subtle Blending of Marrow and Wine

1 **Sautéing marrow.** With a sharp knife, cut marrow out of a cross section of beef marrowbone. Chop the marrow. Heat clarified butter in a saucepan and add finely chopped onions. Add the marrow to the pan and cook it gently with the onions, stirring occasionally, until they are translucent. Add the rice. Stir the grains to coat them with butter.

2 **Pouring in wine.** When the rice has cooked for two to three minutes, pour in dry white wine. Then add a ladle of hot stock—here chicken stock. Dissolve saffron, which is traditionally used in this dish, in 2 tablespoons [30 ml.] of stock and add it to the rice. Stirring, cook the rice over moderately low heat until the liquid is almost absorbed.

3 **Completing the risotto.** Add stock, a little at a time, stirring constantly as each addition is absorbed. Cook the rice until it is just tender *(Step 3, opposite, above).* Take the pan off the heat and stir in butter and freshly grated Parmesan cheese. Serve the risotto with more grated cheese on the side.

2 **Adding stock.** Add a ladleful of heated stock (box, page 25) to the rice, stirring continuously. When the rice has absorbed almost all of the liquid, add another ladleful of hot stock. Stirring all the time, continue to add stock by the ladleful, waiting until each one is absorbed before adding the next.

3 **Sprinkling in cheese.** After about 20 minutes, add stock in smaller amounts and begin testing for doneness. When the kernels are tender to the bite but retain a chewy core, the risotto is ready. Remove the pan from the heat and add chunks of butter and freshly grated Parmesan cheese.

4 **Serving the risotto.** Stir the rice to incorporate the butter and cheese, and season it to taste. Ladle the risotto into soup plates. Serve the risotto at once, while it is still hot and creamy. Accompany the risotto with additional freshly grated Parmesan cheese.

A Colorful Scattering of Peas and Ham

1 **Adding stock.** Sauté chopped ham and onions in hot clarified butter. When the onions are transparent, add rice and stir to coat the grains with fat. Add a ladleful of hot stock to the pan; here, mixed-meat stock was chosen for its hearty flavor. Stirring, add more hot stock as each ladleful is absorbed. Add salt and pepper to taste.

2 **Adding peas.** Have shelled and washed peas ready to add to the rice when it is half-cooked—after 10 to 12 minutes. Gauge the moment for adding the peas according to their age. The firm, mature peas shown here require about 12 minutes to cook. In contrast, tender young peas may cook in as little as five minutes.

3 **Stirring in cheese.** Continue adding stock to the rice, stirring constantly until the rice is almost done; test it as described in Step 3 above. Add chunks of butter. Stir in freshly grated Parmesan cheese. Serve the risotto immediately.

Rice with Its Own Golden Crust

When parboiled rice is simmered gently in a deep pot with fat and a little water, the upper layers steam while the bottom layer fries and crisps. This Middle Eastern technique produces two basic sorts of dishes. If the rice is cooked plain and served with a stew or sauce, the result is called *chelo*. If meat or vegetables are cooked with the rice, as in the demonstration at right *(recipe, page 159)*, the dish is called *polo*.

Traditionally, the rice is soaked before being cooked. Such soaking, regarded as optional in the West, tenderizes the grains and helps the rice cook evenly.

Although soaking partially rehydrates the rice, parboiling in plenty of water must precede the steaming-frying process, because the amount of water used for the second stage of cooking is quite small. In the *polo* demonstrated here, fresh fava beans—also known as broad beans—are parboiled with the rice; other fresh vegetables, such as peas or green peppers, can be treated in the same way. But if you want to include leftover meat in the dish, mix it with the rice after parboiling—together with the flavorings, such as dill, fennel and saffron.

For the second cooking, melted butter is mixed with about half its volume of hot water and added to the drained rice. The butter is essential for the formation of the crust and prevents the rice from scorching. The water evaporates, steaming the upper layers of rice. Too much water, however, would keep the bottom layer of rice from frying.

To brown the crust without burning the rice, the best procedure is to add the buttery mixture to the rice in two stages. Cook over medium heat after the first addition so that the water evaporates and the crust forms, then more gently after the second addition so that the crust does not burn *(Steps 6 and 7)*.

The crust need not be formed of rice alone. For variety, you can mix a small portion of the parboiled rice with beaten eggs, thinly sliced raw potato or sautéed onion; cook this mixture in part of the buttery liquid before adding the rest of the rice and the remaining liquid.

1 **Shelling fresh fava beans.** Put long-grain rice in a bowl, add enough cold water to cover the rice generously, and let it soak for about two hours. While the rice is soaking, split open fresh fava-bean pods with your thumbnail, take out the beans and discard the bean pods.

2 **Peeling the beans.** All but the smallest, bright green fava beans have a tough skin that should be removed before cooking. Pierce each skin with your nail and peel it away.

6 **Starting to form a crust.** In the saucepan used to boil the rice and beans, heat a few tablespoons of water with butter and, if you like, a little saffron. Pour half of the buttery liquid into a bowl and set it aside. Add the rice, beans and dill to the pan and stir lightly with a fork *(above, left)*. With the handle of a wooden spoon, make a deep hole in the mixture *(right)* so that steam can circulate. Cover and cook the *polo* over medium heat for five minutes.

3 **Cutting the dill.** Drain the soaked rice. Wash fresh dill and shake it dry; remove the fronds from the stems and cut the fronds fine. Fresh fennel, treated in the same way, may be substituted for the dill.

4 **Boiling the rice and beans.** Fill a large saucepan almost full with water, add salt and bring the water to a boil. Add the rice and the beans. Boil them vigorously, uncovered, for seven or eight minutes, until the rice is no longer brittle to the bite, but is still firm.

5 **Mixing the dill.** Empty the rice and beans into a colander and rinse them with cold water to stop the cooking. When the water has drained away, put the rice and beans in a bowl and add the cut dill. Use your hands to mix the ingredients together and to separate the grains of rice.

7 **Adding the reserved liquid.** Pour the remainder of the buttery liquid evenly over the *polo*. Cover the pan and let the *polo* cook undisturbed over low heat for approximately 40 minutes.

8 **Serving the polo.** Take the saucepan off the heat and let it cool for five minutes to loosen the crust. Fluff up the steamed rice with a fork, then mound it on a dish. Scrape up the crust and lay it around the steamed rice.

Molding a Dome of Grain

When cooked, sweetened rice is pressed into a mold, then simmered in a hot water bath that is covered to trap the rising steam, the kernels cohere to produce a handsome pudding. The secrets of success in dishes of this type lie in choosing the proper rice and in preparing the mold so that the pudding can be easily and smoothly turned out.

For individual kernels to unite without the addition of binding ingredients such as eggs, the rice must be high in the jelling agent amylopectin: In the heat of the steam, the amylopectin will glue the mass of grain together. The best choice for molding is short-grain rice, preferably the glutinous variety.

Because such rice could easily stick to the mold as well as to itself, the mold should be generously oiled, then lined with slick, oiled plastic wrap *(Step 1, top)*. Other ingredients can be included in the dish to flavor it and also to ease unmolding. For example, in the traditional Chinese rice pudding demonstrated here *(recipe, page 141)*, the prepared mold is partly lined with an arrangement of dried and candied fruits; these serve as a barrier between the rice and the mold—and give the dessert its name— eight-treasure pudding.

Additional ingredients need not be restricted to a barrier role, of course. More dried fruits or stewed fresh fruits could be scattered throughout the mass of rice or tucked into a hollow formed in the center. In this case, the rice mold is filled with a classic Asian confection—a bean purée that is sweetened, enriched with oil, and then cooked until it is reduced to a stiff paste. Red adzuki beans are shown, but mung beans could be substituted to form a green paste.

Garnishes for starchy puddings such as this one should be light. The sauce here is sugar syrup flavored with almond extract and thickened slightly with cornstarch. Fruit-flavored liqueurs or fruit juice could replace the flavoring extract.

1 **Lining the mold.** Line an oiled mold with a large sheet of plastic wrap; oil the wrap. Stuff pitted dates with blanched almonds and cross sections of angelica with candied cherries. Cut slivers and squares from additional angelica and citron. Halve candied cherries. Arrange the fruits close together in the mold, starting at the bottom; they should extend halfway up the sides.

2 **Preparing rice.** Soak glutinous rice overnight, drain it and cook it in a measured amount of water by the Chinese method shown on page 23. Let the rice cool slightly, then stir in sugar. Prepare sweet red bean paste *(box, below)* and let it cool.

Sweetened Bean Paste

1 **Thickening a purée.** Cook soaked adzuki beans until they are tender and purée them in a food processor *(pages 18-19)*. Transfer the purée to a heavy pot, stir in sugar to taste, and cook the purée for about 30 minutes over medium heat, stirring frequently, until the purée thickens and pulls away from the sides of the pot.

2 **Enriching the paste.** Warm vegetable oil in a shallow pan—in this case, a wok—set over medium heat. When a drop of water thrown into the pan sizzles, immediately stir the bean paste into the oil. Continue stirring until the paste has absorbed the oil and become too stiff to stir.

3 **Filling the mold.** Oil your hands to prevent sticking, scoop up a large mass of sweetened rice *(above, left)* and fit it gently into the mold so that it covers the candied fruit completely, holding the design in place. Build a thick layer of rice up around the sides of the mold almost to the rim, leaving a hollow in the center. Spoon the sweet red bean paste into the hollow *(right)* and cover it with a thick layer of rice. Fold the edges of the plastic wrap over the top of the pudding to cover it completely.

4 **Cooking the pudding.** Set a rack in a deep pot, and pour boiling water into the pot to a depth of 2 inches [5 cm.]. Set the molded pudding on the rack. Cover the pot, bring the water back to a boil, and simmer the pudding over medium heat for 45 minutes, replenishing the water if it begins to boil away. Remove the mold from the pot.

5 **Unmolding.** Peel back the plastic wrap to expose the rice, and invert a serving plate over the mold. Invert mold and plate together. Tap the mold all over with a wooden spoon or mallet to loosen the pudding. Gently lift off the mold and peel away the plastic wrap.

6 **Serving.** Stir sugar and water over medium heat until the sugar dissolves. Add almond extract and cornstarch dissolved in cold water; stir until the syrup thickens. Present the syrup in a pitcher and pour it over the pudding at the table.

Couscous: An Ingenious Use of Steam

Conventional techniques for steaming—suspending food over boiling water and cooking it undisturbed in a covered pot—cannot be used for grains: The kernels would heat slowly and unevenly, sticking together in a gummy mass before they softened. However, by modifying the method, tender cracked grains such as millet, wheat or bulgur, or the barley grits used in this demonstration, emerge light and fluffy.

The grain must be thoroughly rinsed and squeezed dry to rid it of surface starch before cooking begins. Then, during steaming, the vessel is left uncovered to allow rising vapors to escape rather than collect and condense on the grain. Even so, the mass of grain must be removed from the steamer regularly and tossed with a fork to break up clumps.

When the water is replaced by a fragrant stew, the steamed grain is deliciously flavored by the stew ingredients and, after cooking, serves as a soft bed for the stew. Innumerable versions of this dish—usually known as couscous—exist throughout Morocco, Tunisia and even Brazil. The stew may be based on beef chuck or round, lamb shoulder, or poultry such as the whole chicken used here (recipe, page 146). The vegetables may range from slow-cooking parsnips, carrots and turnips to fast-cooking summer squash and fresh peas or beans.

No matter what the ingredients, the cooking vessel should consist of a deep bottom section for simmering the stew, and a perforated top section for steaming the grain. Such a vessel can, of course, be improvised with a pot and a cheesecloth-lined colander. But the most convenient vessel is the two-sectioned *couscoussier* shown in this demonstration and available at kitchen-equipment shops.

The stew begins cooking in the bottom of the *couscoussier* before the grain-laden top section is added, because meats and firm vegetables require longer cooking than the grain does. Soft vegetables are added toward the end of cooking (Step 7).

1 **Preparing the stew.** Melt butter in the bottom section of a *couscoussier* and add saffron that has been pounded with turmeric; peeled, seeded and chopped tomatoes; and a halved onion. Swirl the ingredients together, then place a trussed chicken on them. Add a seeded, dried hot chili and a bouquet of parsley and coriander.

5 **Fluffing the grain.** Remove the cheesecloth from the *couscoussier* and lift off the top section. Turn the mass of grain out onto a baking sheet. Break up any lumps in the grain with a fork, then stir in softened butter.

6 **Aerating the grain.** Slowly add water while raking the pieces of grain with a fork to separate them. Smooth the pieces out, then let them dry for 10 minutes. Return the grain to the top section of the *couscoussier* and repeat Steps 4 through 6, but do not add water during the second tossing of the grain.

7 **Augmenting the stew.** Before returning the grain to the *couscoussier* a third time, add fast-cooking vegetables—here, small boiling onions, cylinders of zucchini and yellow squash, fresh peas and fresh fava beans—to the stew. Return the grain to the *couscoussier* (Step 4) and steam for an additional 20 minutes.

2 **Preparing the grain.** Cover the *couscoussier* bottom, set it over medium heat and simmer the ingredients for 15 minutes. Add water and simmer for 30 minutes more. Meanwhile, rinse the grain—in this case, barley grits—in a bowl of cold water. Squeeze the grain, a handful at a time, to extract excess water. Discard the water.

3 **Adding the grain.** Butter the inside of the top section of the *couscoussier*. Add the grain to it, rubbing the pieces between your fingers to break up any clumps caused by the squeezing.

4 **Steaming the grain.** Remove the lid from the bottom of the *couscoussier*. Set the *couscoussier* top section over the simmering stew. Tie a length of dampened cheesecloth around the seam where the two halves of the *couscoussier* meet in order to prevent the escape of steam. Let the grain steam uncovered for 20 minutes.

8 **Serving the couscous.** Remove the steamed grain from the *couscoussier*, turn it out onto a serving plate, and toss it with a little softened butter *(inset)*. Untruss the chicken, place it on the serving platter and surround it with the vegetables. Strain the cooking liquid, mix it with a little cream, and bring it just to a boil to make a sauce. Pour the sauce over the assembled couscous *(right)*.

Grain Soaked to Softness

Steeping grain—that is, pouring boiling water over it and letting it soak up the liquid it needs to swell and soften—plumps the grain without fully gelatinizing its starches. Only a few processed forms are tender enough to be palatable when handled this way: cracked wheat, bulgur, pearl barley, barley grits and steel-cut oatmeal. For this short roster of candidates, the treatment has some notable rewards: The flavor becomes more intense and nutty than it would be if the grain were fully cooked, and the texture becomes pleasantly chewy instead of soft.

Because the kernels do not completely expand in the steeping process, the grain can still absorb more liquid—making it an ideal base for vinaigrette-drenched salads such as the one shown here. In this case, pearl barley and bulgur are paired with sliced nectarine and avocado, whose colors and flavors accent those of the grains; other appropriate fruits include sliced or cubed papaya, mango, pineapple or apple, or peeled segments of orange or grapefruit. Alternatively, the steeped grains could be garnished with fruit and such vegetables as tomato wedges, onion slices, whole snow-pea pods or cucumber strips; or the grains could be garnished with vegetables alone.

1 **Adding water.** Place the grain in a heatproof bowl; in this example, both pearl barley and bulgur are used, each in its own separate bowl. Pour boiling water evenly over the grain, using ½ cup [125 ml.] of water for each cup [¼ liter] of grain.

2 **Draining.** Set the grain aside to soak for half an hour or so, then drain it in a strainer. Return the grain to its bowl, and for each cup of grain, pour in ½ cup of vinaigrette—made with four parts of oil to one part of vinegar. Toss the grain gently with a spoon to coat the kernels. Marinate the grain for five minutes while you prepare the garnishes.

3 **Garnishing and serving.** Halve, pit and slice nectarines. Halve and pit avocados, slip a spoon inside the skin at the broad end of each half, lift out the flesh in one piece and slice it. Brush the fruit with vinaigrette to prevent discoloring. Mound the grain on a serving plate and arrange the fruit around it.

The Secrets of Perfect Porridge

When grits, meals or flaked grains are simmered in water as demonstrated on this page, they yield the porridges, or hot cereals, that are familiar breakfast foods. If the porridges are cooled on a marble slab or in buttered pans, they provide the bases for hearty fried and baked dishes that are appropriate for lunch or supper (*pages 46-47 and 60-61*).

The cooking procedure is straightforward: Salted water is brought to a boil and the grain stirred into it. Because of their small size, grits, meals or flaked grains will absorb enough water to gelatinize within a minute or two. However, for full flavor and body, the porridge should be cooked slowly for an additional 10 to 20 minutes, according to the grain (*pages 14-15*).

While the grain is being added to the water, vigorous stirring is essential to keep the pieces separate and prevent lumps. Whether or not stirring should continue when the porridge is simmering depends on the finished consistency desired. In milling, some of the grain is reduced to tiny particles that become embedded in the surface of the grits, meals or flakes. The particles become pastelike when they gelatinize, and stirring will distribute these sticky particles throughout the porridge. Therefore, producing a porridge with a distinctly granular texture hinges on leaving the grain virtually undisturbed, and stirring it only enough to avoid scorching the bottom. However, for a smoother, more homogeneous texture, frequent—some cooks say constant—stirring is necessary.

In the cooking, the grits, meals or flakes swell to equal the volume of water used. The recommended proportion of water—and of salt—to grain varies with the size of the pieces. As a rule, fine meals require three to four times their volume of water, coarse meals and grits require four to five times, and rolled oats or other flaked grains require three times. And although coarse meal, grits or flakes call for 1 teaspoon [5 ml.] of salt to each cup [¼ liter] of grain, fine meal will need 1½ teaspoons [7 ml.] of salt to each cup.

1 **Adding grain to water.** In a heavy pot, bring a measured amount of water and salt to a boil over medium heat. Then, stirring constantly with a fork, a wood spoon or a whisk (*above*), gradually sprinkle in the grain. In this case, 1 cup [¼ liter] of oatmeal is added to 3 cups [¾ liter] of boiling water.

2 **Cooking the porridge.** Using a wood spoon, continue stirring for about a minute, or until the grain absorbs most of the water and begins to boil: Craters will be formed in the surface by erupting steam (*above, left*). Keep your hand away from the surface to avoid the steam and any flying bits of boiling grain. Reduce the heat to low so that the cereal simmers slowly. Stirring to prevent scorching and adding more boiling water if needed, cook until the cereal is thick enough to almost hold its shape in the spoon (*right*). In this case, the oatmeal simmers about 20 minutes.

3 **Serving.** Spoon the porridge into warm bowls and sprinkle it with sugar—brown sugar is used here. Place a pat of butter in the center of each serving. At the table, pour cream over the porridge.

Grain Steamed in Leafy Packets

Among the more appealing of steamed grain dishes are tidy little parcels formed by enfolding the grains in large leaves; the leaves not only serve as molds but endow the fillings with a delicate fragrance. In China, lotus leaves or palm-like ti leaves are used to enclose fillings based on rice and flavored with such ingredients as sweet bean paste *(pages 30-31)*, shredded meat or water chestnuts. Indonesian cooks wrap mixtures of rice and pork in banana leaves for steaming *(recipe, page 140)*. And Latin America provides the parcels seen here—tamales, made by steaming any of a range of fillings in wrappers of cornhusks, dried beforehand to lessen their strong scent.

Cornhusk wrappers are easy to obtain and easy to prepare for steaming. Dried cornhusks are sold folded together in packets at Latin American markets, or you can prepare your own by removing the silk from fresh husks, separating the leaves, spreading them out in a single layer on baking sheets and drying them in a preheated 200° F. [100° C.] oven for three hours. Either type of dried wrapper always must be soaked in hot water to make it soft and flexible. While the wrappers may be tied around the fillings with cotton string, thin strips of husk are just as effective and much more attractive.

Tamale fillings should be cohesive—to form neat shapes—and usually are highly flavored. Bulk and cohesiveness are provided by corn-flavored grain dough. Latin American cooks use masa—dried hominy ground with water to produce a dough that is ready to use. Masa is available in the United States only at specialty markets in large cities; however, masa harina, or hominy ground without water, may easily be made into a dough, as demonstrated here *(recipes, pages 154-155)*. Most cooks add fire to the bland dough by flavoring the filling with dried hot chilies, which may be chopped or puréed to make a paste. Because dried chilies contain volatile oils just as fresh chilies do, they should be handled with care *(box, page 45)*. In addition to corn and chilies, tamales may hold any ingredients that strike the cook's fancy. Small blocks of cheese, refried beans or shredded meat or chicken coated with thick tomato sauce are all traditional favorites.

Once the ingredients are assembled, the making of tamales is simply a matter of wrapping the ingredients and steaming the parcels. The tamales may be served at once, or cooked ahead of time, then wrapped individually in foil and reheated in a 350° F. [180° C.] oven for 10 to 15 minutes.

4 **Filling the tamales.** Blot a cornhusk dry on a towel and spread it flat. Spread a spoonful of tamale dough in the center of the husk, leaving borders wide enough to cover the package. Press a hollow in the dough and spoon in a little chili paste. Place a rectangle of cheese—in this case, Monterey Jack cheese—in the center of the chili paste.

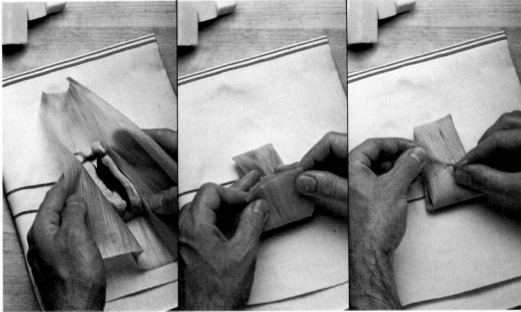

5 **Tying the packages.** Fold the long edges of the cornhusk up and over the filling, squeezing them to make a compact shape *(above, left)*. Overlap the edges on top of the filling, then fold the narrow end and the broad end *(center)* over them. Turn the package seam side down and tie a cornhusk strip around it to secure it *(right)*. Repeat with the remaining cornhusks, reserving some cornhusks for lining the steamer basket.

1 **Cutting strips for tying.** Unfold dried cornhusks, discarding any that are broken. Place the husks in a large bowl and cover them with hot water. Let the husks soak for about one hour, until they are soft and flexible. With a sharp knife, slice one or two husks lengthwise into strips ¼ inch [6 mm.] wide to use as ties for the tamales.

2 **Mixing dough.** Mix masa harina with salt and baking powder. Beat softened lard until it is fluffy. Beat 1 cup [¼ liter] of the masa harina into the lard. Slowly beat in ½ cup [125 ml.] of tepid chicken stock. Continue beating in masa harina and stock until the dough is spongy and a spoonful floats when dropped into cold water.

3 **Making chili paste.** Soak dried hot chilies in water for about 15 minutes, until the chilies are soft. Drain the chilies, slice off the tops and slit each chili lengthwise. Scrape out the seeds and pulp. Force the chilies through a fine-meshed strainer into a bowl, discarding any seeds that remain in the strainer. Wash your hands.

6 **Steaming.** Line a steamer basket or colander with overlapping pieces of cornhusk and pile in the tamales loosely so that steam can circulate. Cover with more husks. Place the basket in a large pot containing a shallow layer of boiling water; the water should not touch the basket. Cover the pot and steam the tamales for about one hour.

7 **Serving.** To test for doneness, unwrap one tamale; the dough should be light and spongy. Let each diner unwrap his tamales, discard the husks and eat the fillings, accompanied, if desired, with a sauce made by flavoring 2 cups [½ liter] of sour cream with ¼ cup [50 ml.] of chili paste.

Frying
Transformations of Taste and Texture

Once softened by cooking in water—or, in a few cases, by only soaking in it—beans, lentils, peas and grains can be fried, an option exploited by creative cooks the world over. In some instances, frying is merely a means of reheating cooked grains or beans in combination with other ingredients. This is the case with such Asian stir-fried dishes as fried rice *(pages 40-43)* or the bean-curd assembly shown on the opposite page. In stir frying, the elements are blended quickly together in a little oil over high heat; the object is for each ingredient to display its own distinctive character, the diversity being united by a bland background of rice or bean curd.

More homogeneous assemblies are made by forming grain or bean mixtures into suitable shapes and pan frying them, producing cakes with crisp crusts. For example, cooked beans mashed into lard in a hot skillet will dry and cling together, forming the thick pancake known in Spanish as *frijoles refritos (pages 44-45)*. For a more delicate effect, soaked but uncooked grains and beans can be puréed and fermented, producing light, tart batters. Batters of this type are the bases of the Ethiopian millet pancakes called *injera (recipe, page 154)*, and of India's crepelike rice-and-bean *dosa (pages 48-49)*, often used as wrappers for spicy fillings.

Cohesive grain and bean mixtures can be fried in deep hot oil as well as in shallow pans: They will not disintegrate in the high heat required for the process. Squares of bean curd, their delicate surfaces protected by a cornstarch batter, are good candidates for deep frying *(pages 50-51)*. And dumplings or croquettes, crisp outside and moist within, can be formed either from fermented bean purées, as in India *(pages 54-55)*, or from cooked rice, as in Italy *(pages 56-57)*.

The choice of fat or oil as the cooking medium depends on the type of frying to be done. Butter may be used for frying pancakes but, because it burns at relatively low temperatures, the butter should first be clarified *(page 24)* or combined with oil. For the higher heats of stir frying or deep frying, a mild-flavored vegetable oil—corn or peanut oil, for instance—is required. Lard may be used for preparing any fried dish, provided that its distinctive sweet taste does not overwhelm the flavors of the other ingredients.

A colorful mélange of snowy bean-curd chunks, whole snow peas and sliced ham is tossed with stir-fried scallions and Chinese mushrooms. Quick frying crisps the flavorings; the bean curd provides a soft contrast.

Stir Frying: The Speediest Method

Stir frying—tossing foods in hot oil just long enough to bring out their flavors and fragrances—is a fast, effective way to marry cooked white rice with other ingredients such as meats, seafoods and vegetables. The dishes made this way are startling in their diversity, but the principles of forming them do not change.

Because stir frying must proceed with all possible speed, the ingredients must be assembled and prepared in advance. Choose long-grain rice—less sticky than medium- or short-grain varieties—and cook it in unlimited water, as shown on pages 22-23. Refrigerate the rice for three or four hours to firm the kernels. As a further precaution against stickiness, separate the chilled kernels (Step 2, right). Other ingredients, such as the garlic, onions, shrimp and pork used here to make Malaysian fried rice (recipe, page 133), should be cut into small pieces that will cook quickly and evenly. To ensure that delicate foods do not overcook

in the commotion of frying, you can precook them and add them last.

Garnishes, too, should be prepared in advance: Stir-fried dishes must be served immediately if they are to retain their fresh flavors and textures. In this demonstration, strips of omelet and slivers of scallion are readied before the stir frying begins, then used to decorate the completed dish on its serving platter.

Stir frying is done in a broad pan such as a skillet or the wok shown here. Use mild-flavored peanut or corn oil as the frying medium; butter would burn, and strong-flavored olive oil would mask the varied tastes of the ingredients. Both pan and oil should be preheated so that the foods sear quickly. Add ingredients to the pan according to the cooking times they require. Here, for instance, stir frying begins with flavorings, thus allowing full release of their essences. Meat comes next and, finally, the elements that require only heating—shrimp and rice.

1 **Preparing the ingredients.** Chop onions and garlic, and reserve them in a small bowl. Slice boneless pork into ⅓-inch [1-cm.] cubes and set them aside in a separate bowl. Parboil shrimp for one to two minutes in salted water and drain them. Peel each shrimp, cut it in half lengthwise, and pull out its dark intestinal vein.

4 **Stir frying onions and pork.** Set a wok over high heat; after 30 seconds, add peanut oil and let it heat for 30 seconds, tilting the wok to coat the surface. Add the chopped onions and garlic, and use a spatula to toss them briskly for about a minute. Add the pork and toss it until it loses its pink color—about four minutes. Add soy sauce and, if you like, dried hot chilies.

5 **Adding shrimp and rice.** Stir in the halved shrimp and cook them for a minute, tossing the ingredients constantly. Then add the rice to the wok and toss it lightly with the other ingredients for about one minute, or until the rice is hot. Immediately transfer the mixture to a warmed serving platter.

2 **Preparing the rice.** Spread cold cooked rice on a tray or large platter. Gently separate any lumps of rice into individual grains, moistening your hands periodically to aid the process.

3 **Preparing garnishes.** Heat peanut oil in a skillet over high heat. When a haze forms on the oil, pour in beaten eggs seasoned with salt and pepper. As soon as the eggs begin to set on the underside, lift the omelet and tilt the pan so the uncooked egg on top runs under the cooked egg on the bottom *(above, left)*. After about one minute, when the omelet is firm but still moist, use a spatula to fold it into a roll *(right)*. Set the omelet aside and prepare any other garnishes—in this case, slivered scallions.

6 **Garnishing and serving.** Spread the slivered scallions over the top of the fried rice, and use the dried hot chilies to decorate the edge of the dish. Thinly slice the rolled omelet crosswise, and sprinkle these slices attractively over the top of the dish.

Binding Agents for Diverse Elements

Some stir-fried rice dishes are made homogeneous and moist by the addition of liquids—marinades, thin sauces, barely cooked eggs or, as in this demonstration *(recipe, page 134)*, a mixture of several of these elements. As with any stir-fried assemblage, careful advance preparation is essential: Cooked rice must be separated into grains; flavorings, vegetables and meats cut up; and meats marinated as required. Here, slices of beef are soaked in a mixture of soy sauce, oil and garlic; because the marinade will double as a sauce for the finished dish, a little cornstarch is added to it so that it will thicken during cooking.

The ingredients are stir fried in the usual way—in sequence according to the cooking time they require. If you do not have a sloping-sided wok, which permits easy tossing, use a skillet. If you do so, clear a space in the pan when you add each new ingredient so that it has a moment in contact with the hot pan.

1 Preparing ingredients. Halve a green pepper lengthwise; seed, derib and sliver it. Slice scallions. Peel fresh ginger root and sliver it; chop a garlic clove. Stem coriander leaves. Cut sliced ham and bacon into squares. Firm a piece of beef tenderloin in a freezer for 30 minutes, then slice it thin.

2 Flavoring the beef. In a small bowl, stir the chopped garlic together with oil, soy sauce, sugar, salt and pepper. Stir in cornstarch. Add the beef slices and coat them well with the sauce. Let the beef marinate in the sauce for a few minutes while you prepare the rice *(Step 2, page 41)*.

6 Breaking eggs into the pan. Again, make a space in the middle of the mixture and break eggs into it *(above, left)*. Season them lightly with salt and pepper. Stir the eggs until they begin to cook but are still very creamy *(center)*, then combine them with the other ingredients *(right)*. Taste the mixture for seasoning and, if you wish, sprinkle a little soy sauce over it for extra flavor.

3 **Starting to stir fry.** Heat oil in a large skillet. Add the ginger to flavor the oil, then drop in the pepper strips. Stir fry them over high heat for about one minute, tossing them constantly with a spatula. Push the peppers to one side and add the rice to the pan.

4 **Adding the ham.** Stirring gently, fry the rice until the grains are hot and separate. Push the rice and peppers to the edges of the skillet, making space in the center. Fry the bacon and the scallions in the space. After a few seconds, add the ham.

5 **Adding the coriander.** Stir the mixture, then make space in the center of the skillet and put in the beef with its sauce. Fry the beef slices for a few seconds, turning them to sear all their surfaces. Add the coriander leaves and stir fry the mixture for a few minutes.

7 **Serving.** Transfer the fried rice to a warmed serving dish (above). If you like, garnish the dish with more coriander leaves. Serve immediately, while the egg is soft and the rice is lightly bound.

Rustic Bean Pancakes

When cooked beans are puréed and fried, the starch they release becomes sticky so that the beans cohere to produce a thick pancake, known in Mexico as *frijoles refritos,* or "well-fried beans." This simple frying technique can be used to produce pancakes of different tastes and textures. Their characters are determined not only by the beans that are used, but also by the methods selected for boiling and puréeing them, the ingredients fried with them, their frying time and the garnishes chosen to accompany them.

Any whole beans—black beans, kidney beans or pinto beans, for instance—can be treated this way. The preliminary soaking and cooking proceed as demonstrated on pages 16-17. The beans can be simmered in water or stock, along with whatever flavorings may appeal—a ham hock, perhaps, or aromatic vegetables. Some Mexican cooks believe that the beans have the best flavor when they are cooked several days before they are to be fried, then left in the refrigerator to steep in their cooking liquid.

Before frying, the cooked beans may be puréed in a processor or food mill as demonstrated on pages 18-19; the pancake formed from them will then be quite smooth in texture. Or you can make a rougher purée, as in this demonstration, simply by mashing the beans as you fry them. Whatever the method chosen, the beans should be puréed with some of their cooking liquid so that they will be moist enough to cohere as a pancake.

The beans should be fried in a large, shallow skillet, which allows for quick evaporation of the liquid. Use oil as a frying medium or, for a rich, distinctive flavor, use lard. If you like, you can add other ingredients to the purée—chopped onions, peppers or bacon, for example.

For a moist pancake, fry the beans only until they cling loosely together. Or continue frying, as here, until the beans are dry and cohesive enough to be flipped over and rolled. In either case, garnish the pancake to provide contrasts of color and texture. Here, shredded lettuce and a pungent raw-tomato sauce known as *salsa cruda (box, opposite)* accompany the beans. Or you could use shredded cheese such as Monterey Jack or Cheddar, radish roses, or sliced scallions.

1 **Frying the onion.** Simmer soaked beans—in this case, red kidney beans—until they are soft *(pages 16-17).* Chop an onion fine. Melt a little lard in a large skillet. Fry the onion over low heat for about 10 minutes, until it is soft. Add a ladleful or two of the beans and their cooking liquid to the onion.

2 **Mashing the beans.** Increase the heat until the liquid comes to a boil. Keep the liquid boiling as you mash the beans with a pestle. Continue mashing until the beans form a coarse purée.

3 **Adding more beans.** Add another ladleful of beans along with their cooking liquid *(above).* Mash them thoroughly into the purée. Continue adding and mashing the beans a ladleful at a time until they are all coarsely puréed. You can either serve the purée at this stage or continue to cook the beans until they dry and form a thick pancake *(Step 4).*

A Piquant Sauce of Uncooked Tomatoes

The tangy Mexican tomato sauce that is called *salsa cruda* ("uncooked sauce" in Spanish) is simplicity itself to make, and its rough texture and sharp flavors add interest to soft, bland bean and grain dishes. The sauce consists of raw ripe tomatoes, onions, garlic and fresh hot chilies, chopped and mixed together with chopped fresh coriander, which Mexican cooks call *cilantro*.

The chilies—in this case, serrano chilies—must be handled very carefully: They contain volatile oils that can irritate skin and eyes. Always rinse chilies in cold water before removing their stems and seeds, and avoid touching your face and eyes while you prepare them. Afterward, be certain to wash your hands thoroughly. If you like, you can attenuate the fire of the chopped chilies by soaking them for an hour in cold, salted water.

1 **Seeding fresh chilies.** In a bowl, combine peeled, seeded and chopped tomatoes with chopped onions and garlic. Cut the tops off fresh chilies. Split each chili in half lengthwise and remove the seeds and fibers from its interior. Discard the seeds and tops.

2 **Mixing the sauce.** Chop the chilies and coriander and stir both into the tomato mixture. Add sugar, some oil or cold water if the sauce seems dry, and salt and pepper. The sauce is best if served immediately, but will keep for up to three hours at room temperature.

4 **Boiling off the liquid.** Keep the skillet over a fairly high heat, shaking it occasionally, until the purée begins to dry out and sizzle at the edges. Tilt the skillet gently from side to side so that the purée moves in a mass and comes away from the sides of the skillet.

5 **Turning out the beans.** Tilt the skillet so that the beans slide to the side farthest from the handle. Then raise the handle so that the bean pancake rolls out of the skillet onto the center of a serving dish. Neaten the edges of the pancake, if you like, but do not try to reposition the pancake.

6 **Serving the beans.** Garnish the finished purée with shredded lettuce. Surround each helping of beans with some of the lettuce garnish and, if you like, a ring of raw-tomato sauce. Add any other garnishes you wish.

Crisp-surfaced Cakes of Porridge

Porridges of all kinds *(page 34)* will cool into solid sheets that can be cut into small shapes, then fried to produce cakes with crisp, brown surfaces and moist. springy interiors. The cakes can be made from one kind of porridge or, as shown here, two kinds that are layered together. Whether the finished dish is sweet or savory depends on what flavorings are added to the porridge and what sauces or garnishes accompany the cakes.

For sweetness, a porridge can be enhanced by the last-minute addition of such ingredients as raisins, chopped figs or dates, or nuts that have been spread on a baking sheet and then toasted for 10 minutes in a 300° F. [150° C.] oven. For a savory cake, flavor the porridge with grated cheese, with bits of fried bacon, or with chopped-and-sautéed vegetables such as green peppers or cabbage.

Whatever the porridge, it is cooked just as if it were to be eaten hot. If you use two porridges, begin one 10 minutes before the other so that both receive the attention they need during the early cooking

stages. The first porridge will stay warm for at least 10 minutes in a covered pan.

To ensure that the porridge cools properly, pour it onto a cold surface that has been oiled to prevent sticking. A marble pastry slab is ideal; alternatively, use a chilled baking sheet. Once poured, the porridge should be immediately spread out and smoothed. When you layer porridges, pour the second out while the first is still hot; otherwise, the first layer will form a top skin that prevents the second from adhering. When it has set, cut the sheet into rectangles or diamonds with a knife, or into disks with a biscuit cutter.

The porridge cakes can be fried in any fat—oil, butter or lard. In every case the fat should be preheated; otherwise, the cakes will be soggy.

The fried cakes can be complemented by a spectrum of sauces and garnishes. Savory cakes, for instance, make a fine foil for a tomato or meat sauce or a garnish of sautéed mushrooms. Here, nut-flecked cakes are accompanied by ham, sausage and maple syrup.

1 **Adding nuts.** Make two porridges— in this case, one from hominy grits and another from cornmeal—following the instructions on page 34 to ensure that the porridges remain free of lumps. Stir toasted, chopped pecans into the hominy-grits porridge.

3 **Cutting squares.** When the porridge is firm, use a knife or spatula to trim it into a rectangle. Divide the rectangle into equal-sized pieces—in this case, 2½-inch [6-cm.] squares. Clean the blade after each cut to keep the porridge from sticking. Set the squares aside on a baking sheet.

4 **Frying the cakes.** Melt lard in a skillet over medium heat. When the lard sizzles, put in several porridge squares, spaced well apart: If too close, the cakes will steam rather than crisp. Fry the cakes for about 15 minutes, turning them once. When each batch is crisp and brown, drain the cakes on paper towels. You can keep them warm in a preheated 200° F. [100° C.] oven.

2 **Layering the porridges.** Immediately pour the hominy-grits porridge onto a flat, oiled surface *(above, left)* — in this case, a marble slab. Using a long, thin metal spatula, spread the porridge in an even layer that is ¼ to ½ inch [6 mm. to 1 cm.] thick *(center)*. Immediately pour the warm cornmeal porridge over the layer of hominy grits *(right)*, and spread it smooth to make a second layer. Let the porridge layers cool completely — about an hour.

5 **Garnishing and serving.** For an attractive breakfast presentation, prick sausages, fry them and drain them on paper towels; then fry slices of ham in the same pan *(above)*. Place the ham and sausages in the center of a large plate and arrange the sautéed porridge cakes in an overlapping ring around the edge. Just before serving, pour warmed maple syrup over the cakes.

Fermentation to Aerate Fragile Wrappers

A batter as smooth as a flour-based pancake batter can be formed from puréed raw split beans and white rice—and it will have a special virtue: In the proper environment, microorganisms in the ingredients will multiply, converting vegetable starches to acids by fermentation and giving the batter—and the pancakes made from it—an appealing sourness.

Many pancakes of this type appear in Indian cuisine; the variations result from choosing different beans, lentils and flavorings and altering proportions. In this demonstration, split black gram is used *(recipe, page 164)*.

The beans or lentils and the rice should be soaked to soften them. Rice requires about 12 hours of soaking; split beans and lentils require at least four. Once soaked, the ingredients are puréed in a food processor or blender, lightly salted, then allowed to ferment in a warm place for about 12 hours. The ideal temperature for fermentation is about 80° F. [25°

C.]; temperatures below 65° F. [18° C.] will slow the multiplication of the microorganisms, and temperatures higher than 115° F. [45° C.] will kill them.

The treatment of the fermented purées determines the texture of the finished pancakes. A little flour—in this demonstration, rice flour, sold at Asian food markets—will give the batter body; a little baking soda will lighten it.

Adding a small amount of water produces a batter that fries into plump, soft cakes; more diluted batters become thin, crisper cakes. During frying the water turns to steam within the pancakes—thus deactivating the toxic lectins. For this reason, the beans do not require boiling before they are used.

The pancakes can be served flat, with melted butter, or wrapped around vegetable or meat fillings. The filling shown here is a mild one—potatoes, scallions, herbs and spices—that will not mask the pleasantly astringent taste of the cakes.

1 Puréeing beans. Soak rice and beans separately in cold water overnight, then drain them; do not use the quick-soaking method, lest it inhibit fermentation. Purée the beans in a food processor or blender, operating it in short spurts and adding up to ¼ cup [50 ml.] of water for each cup [¼ liter] of soaked beans to aid the puréeing.

4 Completing the batter. After 12 hours, when the purée is light and frothy, add a pinch of baking soda and whisk in flour—here, rice flour. Then dilute the batter with water. For thick pancakes, add enough water to make the batter the consistency of lightly whipped cream. For thin pancakes, add water until the batter is like light cream.

5 Cooking the pancakes. Heat a small, heavy skillet or crepe pan and add a little vegetable oil. Pour in a ladleful of the batter *(above, left)* and quickly tilt the pan to distribute the batter evenly over the pan bottom. Cook for about three minutes. Using a large spatula *(right)*, carefully flip the pancake. Cook the second side for about one minute before removing the pancake to a plate. If you prefer very crisp cakes, cook and serve the pancakes individually, oiling the pan for each one. Or continue to cook the pancakes and stack them; they will soften as they wait.

2 **Mixing rice and beans.** Scrape the puréed beans into a bowl and grind the rice in the processor or blender. Add the rice and a pinch of salt to the beans *(above)*. If you did not add water when puréeing, add some now, allowing about ¼ cup for each cup of purée. Cover the bowl and set it in a warm place to ferment for 12 hours.

3 **Making a filling.** In a large skillet, simmer scallions in water until they are soft. Add peeled, cubed potatoes and then stir in spices and herbs *(above, left)*—paprika, black mustard seeds, ground cumin and finely cut chives are used here. Simmer the mixture, covered, until the potatoes are soft—about 15 minutes. Uncover the pan, add chunks of butter *(right)* and toss the ingredients over medium heat until the water has evaporated.

6 **Filling and serving.** Spoon the potato filling across the center of each pancake *(inset)*; fold the edges of the pancake over the filling. Transfer the stuffed pancakes to a serving dish. If you like, garnish the dish with lime wedges and pour melted butter over the pancakes just before serving *(right)*

Enhancing the Delicacy of Bean Curd

The delicate flavor and creamy consistency of bean-curd cakes *(pages 20-21)* provide an excellent foil for the crisp textures and intense flavors of Asian stir-fried dishes *(recipes, pages 108-110)*. Two strategies allow the cook to keep the tender cakes from disintegrating in the high heat and rapid motion of the frying.

One approach, seen in the top demonstration at right, involves simmering the bean curd rather than frying it. The first step is to stir fry the flavoring ingredients—in this case, scallions and dried Chinese mushrooms. Next, the flavored stock used for Chinese white sauce *(recipe, page 167)* is added to the pan and slices of bean curd are briefly and gently heated, along with such accompaniments as the ham and snow peas used here. After the simmered ingredients are removed from the pan, the stock is thickened with cornstarch.

Cooked in this fashion, the bean curd remains velvet soft while absorbing flavor from the stock. Cubed bean curd can, however, be deep fried so that it acquires a crisp surface but stays tender and moist within *(bottom demonstration)*. To get this effect, the bean-curd cubes are coated with a batter that not only protects them from the heat but also cooks during frying to form a golden crust.

For the bean-curd cubes to fry successfully, the temperature of the oil should be about 375° F. [190° C.]. If the oil is too cool, the bean curd absorbs it and becomes greasy; overly hot oil would burn the batter. If you use a deep pan, you can check the temperature with a deep-frying thermometer. If you use a sloping-sided wok, you can test the temperature with a bread cube instead. The oil should have as subtle a taste as the bean curd; peanut oil is traditional.

Deep-fried bean curd can be mixed with almost any combination of ingredients. In this demonstration, the bean curd serves as a backdrop for a spicy dish of stir-fried pork, scallions and hot chilies coated with Chinese white sauce.

Moist Slices to Complement a Stir-fried Assembly

1 **Assembling ingredients.** Soak dried Chinese mushrooms in hot water for 30 minutes. Meanwhile, trim snow peas, chop scallions and dice ham fat. Cut ham into broad slices ¼ inch [6 mm.] thick, and slice bean curd ⅜ inch [9 mm.] thick. Peel and chop fresh ginger. Drain the mushrooms, discard their stems and chop the caps.

2 **Frying flavorings.** Heat a wok, add oil, and tip the wok to spread the oil. Add the ham fat and cook until it is crisp. Add the mushrooms and the white parts of the scallions; stir fry over medium heat for approximately one minute. Add flavored chicken stock and simmer for one minute.

Deep-fried Morsels with an Unexpected Crunch

1 **Coating the bean curd.** Cut bean-curd cakes into large cubes. Prepare a cornstarch-and-water solution that is thick enough to cling to the cubes, but still fluid — ⅓ cup [75 ml.] of cornstarch in ½ cup [125 ml.] of water is about right. Turn half of the cubes in the cornstarch mixture to coat them.

2 **Deep frying the cubes.** Heat a wok and pour in peanut oil to a depth of 3 inches [8 cm.]. When the oil is hot enough, it will brown a cube of bread in 60 seconds. Fry the batter-coated bean-curd cubes for three or four minutes, turning them to brown them evenly. Drain the cubes in a strainer. Coat and deep fry the remaining cubes

3 **Adding other ingredients.** Add the snow peas, ham slices and bean curd to the wok. Without stirring, simmer the ingredients for five minutes; monitor the heat to prevent the stock from boiling and breaking the bean curd. With a perforated spoon or skimmer, transfer the snow peas, ham and bean curd to a plate.

4 **Serving.** On a platter, arrange bean curd, snow peas and ham alternately for a venetian-blind effect *(inset)*. Bring the stock to a boil; add the ginger and the green parts of the scallions. Then gradually stir in a cornstarch-and-water solution to thicken the stock. Pour the sauce over the arrangement *(above)* and serve.

3 **Cooking pork and chilies.** Arrange the fried bean curd on a serving plate and keep it warm in a preheated 200° F. [100° C.] oven. Slice dried red and fresh green chilies crosswise *(page 45)*, slice scallions diagonally, and dice fresh pork. Pour most of the oil out of the wok, and stir fry the pork for four minutes. Add the sliced chilies and cook until the red chilies darken; add the scallions and stir briefly. Increase the heat and toss the ingredients together.

4 **Serving.** To make the Chinese white sauce, add stock to the wok and bring it to a boil. Gradually stir in a cornstarch-and-water solution to thicken the sauce until it coats a spoon lightly. To serve, pour the sauce-enriched pork mixture over the bean curd.

Deep-fried Crusts That Hiss and Crackle

A sheet of white rice, dried over low heat, solidifies into a lacy crust that, after deep frying, gives audible distinction to a range of Chinese dishes: When a moist topping is poured over the fried crust, it produces the hissing sound that accounts for the name "sizzling rice."

Preparation of the crust begins with a seemingly wasteful act: After a potful of rice is cooked, most of the rice is removed, leaving a thin shell at the bottom. This shell is heated until crusty and cohesive. It is then taken from the pot and dried at room temperature for at least a day; tightly wrapped, it will keep indefinitely.

To puff and color it, the crust is broken into pieces and briefly deep fried, following the procedure demonstrated on pages 50-51. The only challenge of the frying is to coordinate it with the cooking of the dish that accompanies the rice crust: Both must be piping hot if the rice and liquid are to sizzle at serving time.

If you intend to serve the crust with a soup that need not be eaten as soon as it is made, cook the soup (recipe, page 136), then keep it hot while you fry the crust. If the crust will be served with sauce-coated stir-fried foods, the cooking order must be reversed, since a stir-fried assembly has to be served the instant it is finished. Deep fry the pieces of crust, then keep them hot in the oven while you do the stir frying.

The assembly shown here, a mixture of shellfish and vegetables (recipe, page 135), follows the fundamental rules for stir frying described on pages 40-41, with some elaborations. To protect its delicate flesh and give it the smooth finish called "velvet" in China, the shellfish is coated with egg white and cornstarch, then cooked briefly to seal the coating. When the shellfish is stir fried with other ingredients later, it steams inside the coating, emerging moist and tender.

To provide the sauce that produces the sizzle, stock and cornstarch are added to the vegetables either before the shellfish or after all the ingredients are stir fried.

1 **Preparing a shell.** Cook long-, medium- or short-grain white rice in measured water in a heavy pot, using the Chinese method described on pages 22-23. When the rice is tender, gently scoop it out of the pot, leaving a ¼-inch [6-mm.] layer in the bottom and around the sides. Reserve the scooped-out rice for use at another time.

5 **Sealing the coating.** Pour 3 inches [8 cm.] of oil into a hot wok, and heat the oil until a cube of bread dropped into it foams slowly. With a skimmer or perforated spoon, lower the shrimp and scallops into the oil, and cook them until their coating turns opaque and sets—about 30 seconds. Remove the shrimp and scallops and drain them in a strainer set over a bowl. Strain the oil.

6 **Deep frying rice crust.** Pour the strained oil into the wok to a depth of 2 inches [5 cm.], and heat it until it browns a bread cube in 60 seconds. Break the rice crust into large pieces and fry two or three at a time for 10 seconds, or until they brown and puff slightly. Put them on a heatproof plate, drizzle hot oil over them, and keep them in a 475° F. [250° C.] oven. Heat a metal platter in the oven.

2 **Removing the shell.** Cook the layer of rice in the uncovered pot over low heat for an hour, or until it colors lightly and pulls away from the pot. Gently lift out the rice crust, using a spatula to free it. Let the crust dry at room temperature for a day before proceeding.

3 **Coating the shellfish.** Peel raw shrimp, cut them almost in half lengthwise, remove the intestinal veins, and spread the halves apart to form butterflies. Cut sea scallops in half. Mix egg white, cornstarch and a little salt with your hand. Add the shrimp and scallops, cover the bowl, and refrigerate it for about half an hour.

4 **Preparing vegetables.** Break broccoli into florets. Trim snow peas. Cut carrots for even cooking by making a series of diagonal slices while rotating each peeled carrot a quarter turn after each cut. Peel and slice fresh water chestnuts. Slice fresh mushrooms.

7 **Stir frying vegetables and seafood.** Pour all but 2 tablespoons [30 ml.] of the oil out of the wok. Add the vegetables to the wok and toss them gently for about two minutes. Increase the heat and add the shellfish to warm it quickly. Add flavored stock and bring it to a boil. Then gradually add a cornstarch-and-water solution to form a Chinese white sauce (recipe, page 167).

8 **Serving the sizzling rice.** Arrange the pieces of rice crust on the heated platter. Transfer the stir-fried mixture to a separate warmed plate. Carry the rice crust and stir-fried mixture to the table, and finish the assembly in front of the diners, pouring the stir-fried mixture over the hot crust to create a sizzling effect.

Zesty Dumplings of Golden Gram

Puréed raw beans form batters that can not only be fried in thin disks to produce pancakes *(pages 48-49)*, but can also be deep fried by rounded spoonfuls, as demonstrated here: The intense heat of the oil will cause the air in the batter to expand and the liquid to turn into steam, puffing the spoonfuls into delectably crisp dumplings. The steam destroys the toxins in the beans—making it unnecessary to boil them for this use.

In Asia, dumplings of this type are usually made from split grams *(recipe, page 95)*; in the Middle East, chick-peas and fava beans are used for similar but somewhat firmer balls *(recipe, page 117)*. Whichever beans, lentils or peas you choose will require preliminary soaking to soften them before they are puréed. To produce delicate dumplings, purée whole beans through a food mill and remove their skins by sieving the purée; for rougher-textured dumplings, purée them in a food processor or blender.

Using a food processor has a time-saving advantage: The processor aerates the purée. If you purée the beans by any other method, they will produce a stiff batter unless they are fermented *(pages 48-49)* for at least two hours to develop bubbles that will cause expansion during cooking. (You can, of course, allow a food-processor-made purée to ferment if you like the tangy taste imparted by fermentation.)

Any purée can be flavored with a range of other ingredients. Here, for instance, the flavorings are fresh spinach, coriander and hot green chilies. Other suitable additions to dumpling purées include small portions of finely chopped cooked meat or shellfish, fresh ginger or onions. A small amount of baking soda or baking powder will give the mixture extra lightness.

The rules for deep frying are the same as those outlined on pages 50-51. Once fried and drained of excess oil, the dumplings can be served as they are or garnished with a light yogurt sauce, as described in the box at right.

1 Puréeing the beans. Soak beans— here, golden gram—in cold water until almost doubled in size; split beans need four hours of soaking and whole ones require 12 hours. Drain the beans and purée them in a food processor or blender, using enough water to create a purée with the consistency of muffin batter. Transfer the purée to a bowl.

2 Preparing flavorings. Top, split and seed fresh, hot green chilies *(box, page 45)*. Wash and dry fresh spinach, strip off the stems and then chop the leaves *(above)*. Chop fresh coriander leaves fine; slice the chilies.

Cooling Sauce Based on Yogurt

In India, creamy chilled yogurt mixtures are served at most meals to provide a counterpoint to the heat of spicy foods such as the dumplings shown at right. Yogurt sauces, known as *raitas* in Hindi, can be enriched or flavored with any of a broad selection of spices, herbs, vegetables or fruits, the choice depending on individual taste and the ingredients at hand. A *raita* might, for instance, contain cooled and chopped cooked carrots or cauliflower; a mixture of cut-up tomatoes, cucumbers and scallions; or even bits of pickled mango or fresh banana.

The sauce that is demonstrated here is made by blending the flavorings and the yogurt in a food processor. Such a mechanically mixed version has a smooth consistency and is thin enough to be poured easily.

Mixing the sauce. Place fresh mint and coriander, grated fresh ginger, seeded fresh chilies, chopped onion, salt, sugar and a little water in a food processor. Blend the ingredients to a fine purée. Add plain yogurt and blend the ingredients briefly. Cover the sauce and chill it for about one hour before serving.

3 **Seasoning the purée.** Sprinkle the spinach, coriander and chilies over the purée, and add baking powder and a little salt. With a rubber spatula, mix the seasonings into the batter.

4 **Frying the dumplings.** Fill a heavy pan with oil to a depth of 2 inches [5 cm.]; heat the oil to 375° F. [190° C.] on a deep-frying thermometer. Drop spoonfuls of batter into the oil *(above, left)*, without crowding them, and fry for four or five minutes. Turn the dumplings over halfway through the cooking to brown evenly. Use a slotted spoon or a skimmer to remove the dumplings; drain them on paper towels. Fry successive batches in the same way.

5 **Serving the dumplings.** Heap the finished dumplings on a warmed platter and, if you like, present them with a bowl of sauce. Here the dumplings are served on a plate spread with chilled mint-coriander sauce.

Grain Croquettes with a Surprise Filling

Among the many transformations possible for leftover cooked white rice are croquettes made by binding the rice with egg, molding it into a ball around a stuffing and deep frying it *(recipe, page 137)*. These delicious morsels are easy to form; a few simple precautions keep the interiors succulent while their surfaces crisp.

To ensure that the rice packages hold together, use rice that has been cooked to a clinging consistency. Leftover risotto *(pages 26-27)*, its natural stickiness augmented with grated cheese, is ideal, but plain boiled or simmered medium- or short-grain rice *(pages 22-23)* can be substituted. If possible, use rice that has been chilled in the refrigerator—well covered—for at least a day: Rice is easiest to mold when it is cold.

The choice in fillings for the rice packages is virtually infinite. You can, for instance, use any thick meat sauce, a chunk of cooked sausage or other cooked meat, or a slice of an easily melted cheese such as Fontina or mozzarella. In this demonstration, mozzarella is combined with ham. The cheese melts during cooking so that, when the croquettes are eaten, the cheese forms the strings that earn this dish its Italian name: *suppli al telefono,* or "telephone-wire croquettes."

Once they are formed, the croquettes require some surface protection from the deep-frying oil. A coat of bread crumbs will serve the purpose; cooking turns the coating into a crunchy outer layer.

As with all deep frying, the oil must be kept at just the right temperature: If it is too hot, the croquettes will burn; if it is too cool, the oil will penetrate the bread crumbs before they crisp and seal, with soggy results. The proper frying temperature is 375° F. [190° C.]; use a deep-frying thermometer to monitor it, or test the oil before cooking by throwing in a small bread cube. If the cube turns brown within one minute, the oil is ready.

Remember that adding the croquettes will reduce the temperature of the oil. To minimize this effect, use a deep pan with enough oil to submerge the croquettes completely, and cook only a few at a time.

1 **Binding the rice.** Put cold risotto or boiled or simmered rice in a large bowl. Reserve a handful of the rice on a plate. Add lightly beaten eggs to the bowl; use your hands to mix the eggs gently with the rice. If necessary, add enough of the reserved rice to make the mixture in the bowl firm.

2 **Cutting up the cheese.** Cut or tear thin slices of ham into small pieces. With a sharp knife, thinly slice mozzarella or a similar melting cheese into strips about 1 inch [2½ cm.] long.

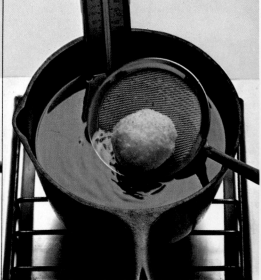

4 **Frying the croquettes.** Pour oil into a large saucepan to a depth of about 3 inches [8 cm.]. Heat the oil to a temperature of 375° F. [190° C.]. Fry two or three croquettes at a time, using a slotted spoon or a skimmer to lower them individually into the oil *(above, left)*. Fry the croquettes until they are golden brown— about one to one and one half minutes. Lift out the croquettes with the spoon or skimmer *(right)* and drain them on paper towels.

3 **Forming the croquettes.** Dust your hands with flour to prevent sticking. Place a large spoonful of the rice mixture on one hand. Fold a piece of ham and lay it on the rice, then add a slice of cheese *(above, left)* and another spoonful of rice *(center)*. Shape the rice into a ball between your palms *(right)*. To coat its surface, roll the croquette on a plate of fine bread crumbs. Firm the croquettes by refrigerating them for about 30 minutes.

5 **Serving the croquettes.** When all of the croquettes have been fried and drained, pile them on a warmed dish *(left)* and serve. Eat them with a knife and fork. As you pull apart the two halves of each croquette, the melted cheese will stretch in strings between them *(inset).*

3
Baking
Exploiting the Oven's Ambient Heat

Grain platforms and containers
Expansion from eggs
Creating sweet gratin toppings
Boston baked beans

By surrounding food with uniform heat, the oven extends a cook's repertoire in many ways. Already-cooked grains and beans, including leftovers, can be married with a variety of other ingredients and rewarmed, without drying out, to form main dishes of unexpected elegance. Raw or parboiled grains and beans can be baked gently to a succulence not obtainable with stove-top cooking methods.

Precooked porridges, for example, can serve as platforms or containers for a spectrum of foods. When cooled, porridges such as the polenta beloved by Italian cooks become firm enough to be sliced into sheets or smaller shapes and layered or covered with meats, sauces or cheese; a brief sojourn in a moderate oven unites the assembled elements into a delicious whole *(pages 60-61)*. Cooked rice—including white, brown and wild varieties—can be molded to make a shell for meats and vegetables; baking endows the rice with a delectable golden crust *(pages 62-63)*.

More homogeneous mixtures—of surprising lightness, considering that they are founded on grains and beans—are made by combining cooked elements with eggs and cream or milk and baking them for an hour or so. Puréed beans mixed with eggs or cream, for instance, become savory custards *(pages 64-65),* fine in texture, yet firm enough to be unmolded, as shown on the opposite page. And grains such as wild rice and buckwheat, enriched with egg yolks and milk, then aerated with beaten egg whites, will puff up into nutty-flavored soufflé-like puddings *(pages 66-67)*.

By contrast, some classic baked dishes begin with uncooked grains or beans rather than with cooked ones and depend for texture and flavor on long, slow baking. Among these dishes is Indian pudding, a combination of cornmeal, milk and molasses. (Its name derives from the fact that early New England settlers, who seem to have invented the dish, referred to maize as Indian corn to distinguish it from wheat, which they called corn.) During its hours in the oven, the cornmeal almost dissolves, giving the pudding a smooth texture unlike that of more briefly cooked porridges. Perhaps the best-loved American dish is Boston baked beans, made, as it was by the Puritans, by combining beans with salt pork and molasses and baking the mixture all day or overnight. The lengthy cooking produces an amalgam with an incomparable richness of flavor.

Leaves of celery-flavored lovage and a gleaming tomato sauce garnish individual black-bean custards. To form the custards, puréed beans were flavored with ground cumin, mixed with eggs and cream, poured into small molds and gently baked until firm enough for unmolding *(pages 64-65)*.

Savory Porridge Casseroles

Grain porridges, formed into cohesive sheets and cut into shapes as demonstrated on pages 46-47, lend themselves to the simplest and swiftest of baking procedures. The porridge pieces can be assembled in layers with complementary ingredients; brief baking blends the flavors of the different elements, turning them into a harmonious whole. A variety of dishes can be made by this method, the final effects depending not only upon the ingredients chosen, but also upon the way they are layered *(recipes, pages 150-151)*. The demonstrations at right show two different uses for polenta—Italian cornmeal porridge (other meals or grits such as buckwheat, hominy or barley could be substituted for the cornmeal).

For the dish demonstrated at top, a sheet of cooled porridge is cut into layers just the size of a baking dish, then stacked in layers with a filling between them. The filling used here is shredded Fontina cheese, whose faintly nutty flavor serves as a fine complement to the cornmeal, but this is only one possibility. The Fontina can be replaced by other firm cheeses such as Emmentaler, Cheddar or Gruyère. Or it can be combined with cooked sausages, sliced hard-boiled eggs or tomatoes, or any other ingredient that suits the cook's fancy.

Baking will heat the filling through and, if it is cheese, melt it to a creamy consistency. At the same time, the porridge layer that covers the assembly—itself sprinkled with Parmesan cheese and melted butter—will crisp and brown in the dry oven heat, to provide a pleasing contrast to the smooth filling within.

In the bottom demonstration, the porridge is cut into small shapes—allowing a more thorough dispersal of other ingredients—then thickly covered with sauce. Tomato sauce *(recipe, page 167)* is shown here, but it could be replaced by meat sauce or vegetables stewed in butter. Baking heats the sauce and porridge together, producing a dish that is soft and moist throughout.

Sandwiching Layers of Cheese

1 **Forming layers.** Shred a generous amount of cheese; Fontina is used here. Make stiff polenta and spread it in a large sheet about ¼ inch [6 mm.] thick to cool for about one hour. Butter a baking dish. When the polenta is firm, use a ruler and a sharp knife to trim it into several pieces the shape and size of the baking dish. Carefully lay a sheet of polenta in the dish *(above, left)*; sprinkle a layer of the shredded cheese over it *(right)*.

Swathing Small Shapes with Sauce

1 **Making sauce.** Place chopped fresh tomatoes, thyme, a bay leaf and a crushed garlic clove in a pan. Set the pan over medium heat, bring the mixture to a boil, and boil it for about 10 minutes, crushing the tomatoes with the back of a wooden spoon. When the tomatoes have softened enough to crush easily, force the mixture through a sieve and set it aside.

2 **Shaping polenta.** Make thick polenta and spread it into a layer about ½ inch [1 cm.] thick to cool for about one hour. When the polenta is firm, cut it into rounds with a cookie cutter. Arrange the rounds in overlapping layers in a buttered baking dish and sprinkle them with a little grated Parmesan.

2 **Finishing the assembly.** Layer polenta and cheese until the baking dish is almost full, ending with a layer of polenta. Sprinkle the top of the assembly with a little grated cheese—in this case, Parmesan—to aid browning, and pour on a little melted butter to prevent drying.

3 **Baking.** Place the assembly in a preheated 400° F. [200° C.] oven to cook for about 30 minutes; check frequently to make sure the polenta is not turning dark brown. If it is, cover with foil. At the end of the cooking time, slide a metal spatula between the polenta layers: The cheese should be creamy.

4 **Serving.** Immediately remove the assembly from the oven and serve it. The cheese will become rubbery if it is overcooked or if it is allowed to set. To serve the baked polenta, cut it into squares with a sharp knife. Using a spatula, carefully transfer the squares to a plate.

3 **Baking.** Stir chopped fresh parsley and basil into the puréed tomato sauce and ladle the sauce over the polenta rounds, thickly coating them *(inset)*. Sprinkle the sauce with more grated Parmesan, and bake the dish in a preheated 400° F. [200° C.] oven for about 20 minutes, until the polenta and sauce are heated through and the cheese melts *(right)*.

A Filled Case of Molded Grain

Bland in flavor and glutinous in character, cooked rice can be molded into a sturdy shell for savory fillings: Surface starch, with a little assistance from eggs stirred in for binding, will make the rice grains cohere. Baking the filled shell in a hot oven will give it a crisp, golden finish, on display when the assembly is unmolded for serving.

Any rice is suitable for this treatment. The short- and medium-grain white varieties, high in surface starch, cohere most easily, but long-grain white rice, brown rice or wild rice may also be used. Although fully cooked leftover rice can be molded and baked, it must be moistened with stock, milk or melted butter to prevent it from drying out in the oven. If you prepare rice just for molding, cook it only enough to soften it partially—about 10 minutes for white rice, 30 minutes or so for brown or wild rice. The rice will finish cooking in the oven, remaining moist and tender under its surface crust.

A bowl or drum-shaped metal vessel such as the charlotte mold used in this demonstration is best for shaping the rice case: The curving sides ensure easy unmolding and the metal conducts heat efficiently to crisp the surface of the rice. To encourage a crust to form, butter the vessel and coat it with bread crumbs.

About 6 cups [1½ liters] of cooked rice is sufficient for lining the 2-quart [2-liter] mold shown here. The layer on the sides and bottom should be about ¾ inch [2 cm.] thick so that the shell holds its shape when unmolded but remains easy to cut. To give the assembly a firm base, cover the lined-and-filled mold with a layer of rice 1 inch [2½ cm.] thick.

Rice molds can be filled with almost any combination of vegetables, meats, sauces and flavorings, precooked to the point where they are moist without being runny: Excess liquid could seep through the rice. The version at right *(recipe, page 137)* contains fennel-flavored sweet Italian sausages braised with onions, tomatoes and meat stock, and is flavored with dried Italian mushrooms.

Garnishes for assemblies of this type also may be varied at will. In this demonstration, barely cooked asparagus spears provide a crisp note; the sauce is reduced meat stock enriched with cream.

1 Mixing the crust. Boil rice *(pages 22-23)*—in this case, short-grain Italian rice—for about 10 minutes, until it is half-done. The grains will still be chewy to the bite. Drain the rice, run cold water over it and drain it again. Transfer the rice to a large bowl. Stir eggs, grated Parmesan, salt, pepper and a little nutmeg into the rice. Set the rice aside.

2 Mixing the filling. Soak dried Italian mushrooms in warm water for 30 minutes; drain and stem them and slice the caps. Strain the soaking liquid and add one third of it to the meat stock; reserve the remaining liquid. Peel, seed and chop tomatoes. Sauté chopped onion in olive oil for 10 minutes. Add peeled, chopped sweet Italian sausages, and continue sautéing for 10 to 15 minutes, until the meat is brown.

5 Making a garnish. Trim asparagus spears and parboil them for two to three minutes. For the sauce, combine meat stock with the remaining mushroom liquid and boil until the mixture reduces by two thirds. Add heavy cream and boil until the mixture lightly coats a metal spoon. Keep this sauce warm.

6 Unmolding. After 25 to 30 minutes of baking, check the rice; it should be golden and should have shrunk somewhat from the sides of the mold. Remove the mold from the oven and let it cool and settle for about 15 minutes. Then run a knife around the edges of the mold to loosen the rice casing.

3 **Finishing the filling.** Stir the prepared tomatoes into the sausage mixture and simmer the mixture for about 15 minutes, until it is thick but not dry. Ladle flavored stock into the pan and simmer, uncovered, for 30 minutes more, until the liquid has evaporated. Stir in the mushrooms and transfer the filling to a bowl to cool.

4 **Lining a mold.** Butter a metal mold and coat it with fresh bread crumbs. Cover the bottom of the mold with the cooled rice and tamp the rice down to make a layer about ¾ inch [2 cm.] thick *(above, left)*. Build up a ¾-inch layer of rice around the sides of the mold, stopping about 1 inch [2½ cm.] from the rim. Spoon the sausage filling into the hollow formed by the rice *(right)*, and cover the filling and rice border with a layer of rice about 1 inch thick, pressing the rice gently to make an even, compact layer. Set the mold in a preheated 400° F. [200° C.] oven.

7 **Serving.** Invert a heated serving platter over the top of the mold, then invert the platter and mold together. Lift the mold from the rice assembly. Surround the rice with the asparagus spears and spoon on a little of the sauce. Serve the rice mold in wedges, passing the remaining sauce separately.

Velvety Custards Fashioned from Beans and Peas

Dried beans produce a surprisingly delicate dish when they are puréed, lightened with cream, bound with eggs and baked. The result is a savory custard, fine in texture but firm enough to be unmolded. Any whole bean, lentil or pea can be treated this way, and the requisite procedures are simple.

The initial cooking of the beans, lentils or peas proceeds as shown on pages 16-17, but for extra flavor they can be cooked in meat stock *(recipe, page 166)* and flavored with herbs: Sage and tarragon will enhance most beans; garlic is a good addition to Great Northern beans or chickpeas; and thyme is well-suited to black beans and lentils.

To ensure lightness, the cooked beans should be puréed *(pages 18-19)* with a little of their drained cooking liquid: ¾ cup [175 ml.] of liquid for every 2 cups [½ liter] of beans. To enrich the purées and make them firm enough to unmold, beat 2 tablespoons [30 ml.] of heavy cream and two eggs into every 2 cups of purée.

The purées can be baked in individual molds *(page 58),* or they can be cooked in larger molds, such as the loaf dish shown here. And you can use either a single puréed bean for a custard or, for greater complexity of taste and a striped effect, you can bake layers of complementary purées, as in this demonstration.

No matter what the custard's size, the cooking heat must be kept gentle and even—350° F. [180° C.] is ideal—to prevent the eggs from toughening as they set during baking. Putting the molds in a heat-diffusing water bath ensures that the proper temperature is maintained during cooking. The 1½-quart [1½-liter] custard demonstrated at right requires about an hour of baking. Small, ½-cup [125-ml.] custards need 15 to 20 minutes.

Bean, lentil or pea custards can be served hot or at room temperature. Garnishes should be chosen to complement the flavors and colors of the elements. A tomato sauce *(recipe, page 167)* contributes a sweet note to the hearty bean purée on page 58, for instance. The less assertive layered custard here is better partnered with an herbed mayonnaise *(box, right; recipe, page 167).*

1 Binding the purées. Separately cook white Great Northern beans, green peas and red kidney beans in meat stock with aromatics—onions, carrots, celery and a bouquet garni. Discard the aromatics and purée the beans and peas separately with some of their cooking liquid. Into each purée whisk eggs and heavy cream; add salt to taste.

2 Making the first layer. Butter a baking dish. Trim parchment paper to fit the bottom of the dish, butter the paper on both sides and fit it into the dish. Butter a second piece of paper—cut to fit the top of the dish—on one side and reserve it. Spoon one purée—in this case, the kidney bean purée—into the dish and smooth it into an even layer.

How to Make Herbed Mayonnaise

Forming an emulsion. Purée blanched leaves and herbs—here, spinach, tarragon, watercress and parsley—in a food processor. Whisk two room-temperature egg yolks with a little lemon juice until they are foamy; whisk in the leaves and herbs *(above, left).* Whisking steadily, add olive oil by drops; allow about 1 cup [¼ liter] of oil for the two yolks. When about half of the oil has been added, increase the flow to a thin stream *(right),* whisking until the mayonnaise is thick but pourable. Add salt and pepper.

3 **Filling the dish.** Let the layer of purée settle for 10 minutes. Then—working across the layer without disturbing the purée already in place—carefully spoon on a second purée—the Great Northern bean purée, in this instance. Wait 10 minutes, then repeat the process with the third purée.

4 **Baking.** Cover the top layer of purée with the buttered parchment paper, buttered side down. Set the baking dish in a larger dish or pan; place this assembly on the middle shelf of a preheated 350° F. [180° C.] oven. Pour enough hot water into the large dish to come halfway up the sides of the baking dish.

5 **Testing for doneness.** Bake the custard for one hour. Then lift the parchment-paper lid and insert a skewer into the center of the custard; if the skewer comes out clean, the custard is done. If not, bake it for five minutes more and test again.

6 **Unmolding.** Remove the custard from the oven and let it set for 15 minutes. Run the tip of a knife around the warm custard to free the top edge from the baking dish. Invert a serving platter onto the dish, then invert the platter and dish together. Lift the baking dish from the custard. Peel off the paper lining.

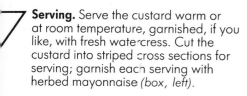

7 **Serving.** Serve the custard warm or at room temperature, garnished, if you like, with fresh watercress. Cut the custard into striped cross sections for serving; garnish each serving with herbed mayonnaise (box, left).

Grain Pudding Lightened with Egg Whites

Egg whites offer distinctive benefits to baked grain dishes: When the whites are beaten, they trap air bubbles that expand during baking, lightening and puffing whatever other ingredients they are combined with. Grain dishes that derive their fine textures from beaten egg whites *(recipe, page 152)* are known as soufflé-puddings because they combine the richness of a pudding with the airy qualities of a soufflé.

The bulk of a soufflé-pudding is produced by combining cooked grains with egg yolks and milk, then simmering the mixture gently to reduce it to a thick, custard-like base. In the demonstration at right, 2 cups [½ liter] of cooked wild rice and 2 cups of cooked toasted buckwheat groats *(pages 14-15)* are mixed with six egg yolks and 1½ cups [375 ml.] of milk to form the pudding base.

As long as the proportions remain the same, the ingredients used in a soufflé-pudding base can be altered: Either of the grains chosen here could be replaced by other distinctively flavored varieties such as millet, brown rice or oat groats; or a single grain could be used instead of two. The pudding base also can be enhanced by fresh herbs—tarragon and rosemary are good choices—or by adding about 1 cup [¼ liter] of grated sharp cheese, such as Cheddar or Parmesan.

To give the soufflé-pudding the proper puffiness, be generous when adding egg whites; eight were used in this case. Beat as much air as possible into the whites before adding them to the base. To help incorporate air, the whites should be at room temperature: They are more elastic and therefore can expand more than refrigerated whites. Because even a speck of fat can prevent expansion, the whites must be carefully separated from their fatty yolks, and the bowl and whisk used must be immaculate.

Whites are best beaten in a copper bowl: The metal causes a reaction that strengthens the bubbles and helps them hold air longer. If you lack a copper bowl, you can instead strengthen the bubbles by adding cream of tartar early in the beating process, allowing about ¼ teaspoon [1 ml.] for every four whites.

To minimize the loss of air, beat the whites just before folding them into the pudding base, and be swift but gentle when folding. Bake the soufflé-pudding immediately after the egg whites have been incorporated.

Puddings like this can be baked in a dish of any shape—round, oval or rectangular. A dramatic presentation may be achieved by selecting an undersized container and fitting it with false sides in the form of a paper collar. After baking, the paper collar is removed, leaving the soufflé-pudding puffed high above the rim of the dish.

4 **Lightening with egg whites.** When the grain mixture has cooled, beat egg whites in a separate bowl until they form stiff peaks. Stir a spoonful of the whites into the grain to lighten it, then turn the rest of the whites into the bowl of grain. Working quickly and lightly with a spatula, fold the whites into the grain.

5 **Baking the pudding.** When the grain mixture and whites are evenly combined, scrape them into the prepared baking dish. Set the dish in a preheated 350° F. [180° C.] oven to cook for 45 minutes. At the end of the cooking time, thrust a metal skewer into the center of the pudding. The skewer should come out clean. If it does not, bake the pudding a few minutes more and test again.

1 **Preparing a baking dish.** Wrap a double layer of parchment paper around a baking dish—here, a 5-cup [1 ¼-liter] soufflé dish—so that it extends 2 inches [5 cm.] above the rim; clip the ends of the paper together while you secure it with string. Brush the inside of the dish and paper with melted butter.

2 **Making the pudding base.** Cook toasted buckwheat groats and wild rice separately; mix them together in a large pot. Whisk egg yolks and milk together until the liquid is smooth; stir it into the cooked grains and season the mixture with salt and pepper.

3 **Thickening the pudding base.** Set the pot of grains over low heat. Stirring constantly, cook the mixture until it thickens slightly and loses its soupy appearance—five to 10 minutes. Transfer the mixture to a bowl and allow it to cool to room temperature.

6 **Serving.** As soon as the pudding is done, remove it from the oven. Cut the string that secures the paper collar around the baking dish and remove the collar *(inset)*. Serve the pudding immediately *(right)*.

Old-fashioned Puddings Slowly Cooked

Long, gentle baking in liquid transforms uncooked ground grains such as cornmeal and hominy grits or whole grains such as rice into puddings whose textures are infinitely more smooth and creamy than those achieved by cooking the same grains on top of the stove. Although details of the process may vary, all grains can be treated this way.

The grains are always baked in large amounts of liquid. For the cornmeal pudding at right, the ratio of liquid to ground grain is 8 to 1—as opposed to the usual 4-to-1 ratio in porridge (page 35). Similarly, while rice cooked on top of the stove (pages 22-23) is most often combined with twice its volume of liquid, rice baked in a pudding (below) is cooked in eight times its volume of liquid.

This abundance of liquid is necessary because of the lengthy baking required to break down the grains into a homogeneous mass: A cornmeal pudding may require as much as four hours of baking, a rice pudding may need two. Liquid naturally evaporates during baking, and if the proportion is too low the liquid will evaporate completely before the grain has reached the proper texture, producing a dry, gritty pudding.

The liquids and grains used for these puddings can be augmented in a variety of ways. The mixtures may, for instance, be thickened with egg yolks or whole eggs or enriched with butter or cream. Sweet puddings may contain sugar, maple syrup or molasses; they may be spiced with cinnamon, nutmeg or ginger, or given textural interest by nuts and fruits.

The puddings may also be endowed with different finishes. For example, the cornmeal pudding is covered by a film of milk or cream during the last half of its baking period (recipe, page 149); the oven heat caramelizes the natural sugars in the milk, giving the pudding a rich, brown crust. A simpler approach—used for the rice pudding here (recipe, page 130)—is to cover a finished pudding with a mixture of brown sugar and butter, then briefly place it under a broiler or in a very hot oven to crisp the topping.

Milk for a Caramel Finish

1 **Preparing the porridge.** Bring milk to a boil in a pan set over medium heat and, stirring constantly, gradually pour in cornmeal. Add sugar and, stirring often, simmer the mixture for about 10 minutes, until it forms a thin porridge. Remove the pan from the heat and stir in butter, then molasses and nutmeg. Add a cinnamon stick.

A Crusty Cover of Sugar and Butter

1 **Mixing ingredients.** Heat milk just to a simmer in a saucepan set over medium heat. Remove the pan from the heat and add sugar, stirring until the sugar dissolves. Place rice and a cinnamon stick in a buttered baking dish. Stir in the warm milk and, if you like, dark and golden raisins.

2 **Adding a brown finish.** Set the baking dish in a preheated 325° F. [160° C.] oven, cover, and let the pudding bake for one and one half hours, until it is thick and creamy. Remove the pudding from the oven. Crumble brown sugar and softened butter together and scatter this mixture over the pudding.

3 **Serving the pudding.** Set the baking dish under a preheated broiler for two to three minutes or in a preheated 450° F. [230° C.] oven for 15 minutes. The sugar topping will melt into a rich, brown glaze. Serve the pudding hot or warm.

2 **Beginning the baking.** Generously butter a shallow baking dish. Spoon the cornmeal mixture into the dish; set it in a preheated 400° F. [200° C.] oven.

3 **Stirring the pudding.** After about 10 minutes of baking, the cornmeal mixture should be thick and bubbling. Stir it well to distribute the flavorings. Then reduce the oven temperature to 300° F. [150° C.].

4 **Adding a milk glaze.** Let the pudding bake for one and one half hours. Then gently pour a thin layer of cold milk or cream onto the brown top of the mixture.

5 **Serving the pudding.** After two and one half more hours of baking, the milk layer will have formed a rich, brown glaze; the pudding beneath will remain soft and smooth. Serve the pudding hot or warm, garnished with ice cream or, as here, with lightly whipped heavy cream.

Melding the Flavors of Pork and Beans

Combined with meat and flavorings, and baked for long periods in a very slow oven, beans gain a uniquely full-bodied taste. The low heat keeps the beans from disintegrating during the extensive time needed for a complete exchange of flavors among the ingredients.

Any fleshy beans—Great Northern, navy or kidney beans, for instance—may be baked. They require the usual soaking *(page 16)* and in most cases are partially precooked by parboiling.

The beans are baked with a fatty meat for richness: During cooking, they will absorb rendered fat and flavor. Salt pork is traditional for Boston baked beans *(demonstration, right and below; recipe, page 105)*, but a smoked ham hock or spareribs could be substituted, and some cooks use calf's feet or pork sausages. The pork may be parboiled beforehand to rid it of salt—in which case, salt must be added to the beans. Or the pork may be left uncooked, as here; it will probably provide enough salt to flavor the dish. To

expose meat surfaces for rendering fat and flavor, the salt pork should be scored. And for even distribution of flavor, some of the pork may be cut into cubes and arranged throughout the beans.

Most cooks add to the assemblage's complexity of taste by including aromatic vegetables such as onions, and herbs and spices such as bay leaves, cloves, mustard, ginger and pepper. Generally, molasses provides the smoky sweetness that characterizes Boston baked beans. But the molasses may be supplemented with—or even replaced by—brown sugar, maple sugar or maple syrup.

To minimize evaporation during baking, the vessel used should be narrow-mouthed, deep and lidded. The ingredients should be immersed in liquid during most of the baking period. Toward the end of the cooking time, the vessel is uncovered and the surface of the mixture allowed to dry somewhat so that the beans develop an appealing crust and the meat a rich, brown veneer.

1 **Preparing ingredients.** Soak navy beans overnight. Parboil them for 10 minutes; drain them, reserving the liquid. Assemble flavorings—dry mustard, salt and an onion stuck with cloves. Slice the rind from a large piece of salt pork; cube a slice of the meat. Deeply score the remaining meat lengthwise and crosswise.

3 **Adding liquid.** Place the scored piece of salt pork on top of the onion and add enough beans to come to the level of the salt pork. Spoon molasses over the salt pork, then pour in enough reserved, hot bean cooking liquid just to cover the beans. Cover the pot tightly and place it in a preheated 275° F. [140° C.] oven.

4 **Browning the salt pork.** Bake the beans for seven hours, checking every hour to make sure the top layer is moist. If not, ladle on enough hot bean cooking liquid just to cover the beans. An hour before the end of cooking, remove the lid to let the beans form a crust. The beans will be tender to the bite when they are done.

2 **Layering the ingredients.** Place the salt-pork rind in the bottom of a deep earthenware casserole—a traditional bean pot is used here—and ladle a layer of beans over it. Scatter the cubed salt pork over the beans. Ladle some of the molasses over the salt pork (*above, left*) and sprinkle in a little dry mustard. Spoon on another layer of beans (*right*); embed the onion in them and add more molasses and mustard. Spoon beans around the onion.

5 **Serving.** Remove the salt pork from the bean pot and reserve it. Remove the onion and discard it. Taste the beans and add salt if necessary. To serve each diner, ladle a helping of beans onto a warmed plate, then slice off a helping of salt pork and lay it on top of the beans.

4

Assemblies
International Food for Feasting

The technique of salting meats
Layering meats and beans
Marinating lamb in yogurt
A fiery sauce from beans
Cleaning squid and crabs

Plain, filling staples though they are, beans and grains provide the foundation for a number of elaborate preparations both famous and festive. None of these dishes is particularly difficult to assemble; the techniques required for making them are demonstrated earlier in this volume. Most of the dishes are, however, time-consuming to produce because they are formed by cooking beans or grains with an array of separately prepared ingredients. Each dish lends itself to almost infinite variation.

The cassoulet of southern France *(pages 74-77)* is a case in point. Cassoulet—it takes its name from the vessel traditionally used for baking it, a *cassole d'Issel*—is a stew based on white beans. It includes fresh pork, salt pork or bacon, preserved goose, and often game or lamb as well. All of these elements should be separately prepared, making the assembly of a cassoulet an impressive undertaking indeed—but one with equally impressive results. The French novelist Anatole France described with pleasure a cassoulet that he claimed had been cooking for 20 years, being replenished at intervals, yet always remaining the same basic stew; France compared its rich, subtle flavors to the sight of the amber flesh tints with which Venetian painters endowed their models.

Brazil's national feast, *feijoada completa (pages 78-81)*, is similarly complex and also revered. This stew of black beans, meats and side dishes may take several days to prepare. However, the cornucopia finally presented is so magnificent that the Brazilian composer Heitor Villa-Lobos was moved to write "A Fugue without End" in its honor.

Fabled preparations based on rice are even more numerous than those made from beans, although most are founded on simple pilaf *(pages 24-25)*. The embellishments added to this preparation raise it to an exalted level. For instance, Indian pilafs *(pages 82-83)*—the variations number in the dozens—may include vegetables such as cauliflower or meats such as lamb; their garnishes range from fried parsley to edible silver leaf. And Spain's paella *(pages 84-87)*, which began as a relatively uncomplicated rice dish enriched with snails, eels and green beans, includes, in later versions, clams, shrimp, chicken and even rabbit. The range of ingredients for pilafs and paellas, as well as for dishes built on beans, is unlimited; the only requirement is a generous hand, since all of these are foods for feasting.

Shrimp and clams adorn an elaborate Spanish paella *(pages 84-87)*. Other additions to the rice beneath the shellfish include squid, spicy sausage, rabbit, tomatoes, peppers, artichokes and green beans. The rice itself is scented with olive oil and powdered saffron in the traditional Spanish style.

73

Southern France: The Many-splendored Cassoulet

The apotheosis of baked beans and the pride of Languedoc, a region of southern France, cassoulet appears in more versions than can be enumerated. In addition to the beans and various forms of pork, most cassoulets contain preserved goose—meat that has been salted, then cooked and potted in its own fat. These are the principal ingredients of the cassoulet made in Toulouse. However, the classic cassoulet of Castelnaudary contains no goose, that of Carcassonne contains lamb and sometimes partridge, and other Languedoc towns have their own variations (*recipes, pages 98-99*).

For the American cook, the challenge in creating cassoulet is to achieve the pungent, earthy tastes that characterize the dish, using somewhat different ingredients from those found in France. French recipes, for instance, may call for locally grown white beans. But any white variety, such as Great Northern, can be used. Similarly, preserved goose is available only at specialty food markets in large cities. An efficacious replacement, shown here, is a fresh goose that is salted, left to steep overnight, then cooked in its own fat. Other necessaries, such as salt pork, pig's foot and garlic-flavored poaching sausages, are readily available from good butchers.

Once the ingredients are in hand, producing a cassoulet demands only time: The components of the dish should be separately precooked, then braised together in the oven. The beans, for instance, are simmered with the pork and sausages; gelatin from the pork rind and the pig's foot adds a velvety smoothness to the cooking liquid. Other meats—in this demonstration, lamb—are braised separately with vegetables in some of the liquid from the beans and pork. Nothing is wasted: The braising liquid is used later as a sauce.

Once everything has been precooked, the cassoulet is assembled. Layers of beans are alternated with the meats in a large casserole, and the whole is moistened with the lamb sauce. Bread crumbs form a top layer that crisps during baking. As the cassoulet cooks, its crust is broken and basted: The result is a deep and crusty layer that contrasts with the moist meats and beans beneath.

1 **Salting a goose.** Pull the fat from the goose cavity and refrigerate it. Cut the goose into pieces, rub them with salt and ground herbs, cover, and refrigerate overnight. Then render the fat by cooking it gently with a little water; strain it into a large pot. Dry the goose pieces and put them in the fat.

2 **Cooking the goose.** If there is not enough goose fat to cover the pieces, supplement it with melted lard. Simmer the goose slowly, uncovered, for one and a half to two hours, until it is tender. Remove the goose pieces from the pan. Strain and reserve the fat.

5 **Boiling the beans.** Bring the water to a boil, boil for 10 minutes, then reduce it to a simmer. After 40 minutes, remove the sausages. After a further 20 minutes, or when it is tender, take out the salt pork. Continue to simmer the remaining ingredients for another hour, or until the beans are tender.

6 **Cutting up the pig's foot.** Drain the beans, reserving the liquid. Untie the rind and slice it into small squares. Cut the pig's foot in half lengthwise. Remove the bones and cut the meat into small chunks. Chop the carrots, but discard the onion—it will have yielded its flavor to the cooking liquid—and the bay leaf.

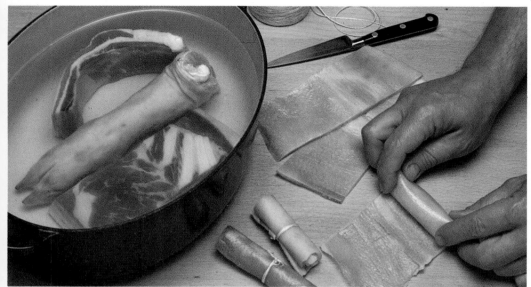

3 **Rolling pork rind.** Slice the rind from large pieces of salt pork. Place a pig's foot and the salt pork in a pot, and cover them with cold water. Cut the salt-pork rind into several rectangles, roll the rectangles tightly and tie them into neat bundles. Add them to the pot. To cleanse the meat of proteins that form a gray scum during cooking, slowly bring the water to a boil and simmer for a few minutes. Drain the meat and rinse it under running water.

4 **Adding beans.** Transfer the meats to a clean pot. Add soaked *(page 16)*, drained Great Northern beans, herbs and vegetables—thyme, bay leaf, garlic, carrots and an onion stuck with cloves. Prick garlic-flavored sausages and add them. Add cold water to cover the ingredients by 2 inches [5 cm.].

7 **Browning lamb.** Chop fresh carrots and onions. Brown them in goose fat, then tip them into a strainer set over a bowl. Reserve the vegetables; when the fat has drained from them, pour it back into the pan. Cut lamb—a shoulder was used here—into large pieces. Put the pieces of meat into the pan, salt them and brown them. Sprinkle on flour *(above)*. After four or five minutes, return the browned carrots and onions to the pan. Add peeled, seeded and chopped tomatoes, garlic and a bouquet garni.

8 **Deglazing the pan.** Pour red or white wine into the sauté pan. Stir gently with a wooden spoon to loosen any scraps of meat and vegetable adhering to the sides and bottom. Add enough of the bean cooking liquid to cover the contents of the pan. Put a lid on the pan and simmer gently for about one and one half hours. Skim off any surface fat two or three times during the cooking period. At the end of cooking, discard the bouquet of herbs. ▶

9 **Making a sauce.** Reserve the lamb and carrots. Tip the remaining contents of the pan into a strainer set over a saucepan. Press the mixture through the strainer with a pestle or spoon. Set the pan half off the heat and simmer the sauce for 15 minutes, removing the fat that collects on the cooler side.

10 **Assembling the first layer.** Rub the inside surface of a large, deep casserole—preferably of earthenware—with cloves of garlic. Scatter the squares of pork rind over the bottom of the casserole. Place the goose pieces on top of them. Spread about a third of the drained beans over the goose.

13 **Adding the sauce.** Without displacing the bread-crumb topping, gently spoon the strained sauce from the lamb stew over the casserole. Add the sauce until liquid reaches the level of the bread crumbs. If there is not enough sauce, supplement it with liquid from the beans and pork.

14 **Adding goose fat.** Dust the casserole with more crumbs. Sprinkle goose fat over the top (above). Place the uncovered casserole in a preheated 425° F. [220° C.] oven. Bake until the casserole is bubbling—about 20 minutes. Then reduce the temperature to 300° F. [150° C.] to maintain a gentle simmer.

15 **Breaking the crust.** Baste the dish every 20 minutes or so with the leftover lamb sauce or, if that is used up, the bean-and-pork liquid or the cassoulet's own liquid. When the crust begins to turn golden, after about one and one half hours, break up the surface with a spoon.

11 **Forming the second layer.** Arrange the cooked lamb and carrots over the first layer of beans *(above)*. Add the pieces of pig's foot and the chopped carrots that cooked with the beans and pork. Cover the meat with about half of the beans remaining in the colander, spreading them evenly.

12 **Adding the third layer.** Slice the garlic sausages and the salt pork into pieces about ½ inch [1 cm.] thick. Scatter the sausage and pork on top of the second layer of beans *(above)*. Ladle the remaining beans over the pork and sausage. Sprinkle the surface with fine bread crumbs.

16 **Forming a perfect crust.** Baste the broken surface of the cassoulet with the bean-and-pork liquid. Bake the dish for one and one half hours more. Break and baste at least four times, until a thick crust forms *(above)*. Serve the cassoulet, making sure each diner receives crust, meat and beans *(right)*.

Brazil: Orchestrating a Feijoada

As it does in France, the humble bean provides the foundation of a celebrated, ceremonial dish in Brazil—the *feijoada completa*. This colorful feast—which derives its name from *feijão,* the Portuguese word for bean—consists of black beans and assorted meats that are simmered together for hours, then served separately in a lavish display that includes several satellite dishes.

Although a *feijoada* is a kind of culinary extravaganza, its success depends more upon careful planning and budgeting of time than on sophisticated cooking skills. The first task is to assemble the formidable list of ingredients that a *feijoada* typically includes. Tradition requires that salt-cured beef—lean beef that has been salted and sun-dried—and smoked tongue be in the stew, and most *feijoadas* include other meats as well. An elaborate *feijoada* such as the one demonstrated at right and below *(recipe, page 190)* may contain corned spareribs,

chorizo (a spicy, smoked pork sausage), fresh pork sausage, a pig's foot, lean bacon and lean beef chuck.

The more unusual meats—salt-cured beef, *chorizo* and corned spareribs, for instance—are sold at Latin American food markets and by some butchers. If necessary, you can substitute: *Chorizo* can be replaced by any smoked sausage, and if corned spareribs are not available, you can give fresh pork spareribs a similar taste by coating them with coarse salt and refrigerating them for five days.

Preparation of the meal itself must begin at least a day in advance: The smoked tongue, corned or salted spareribs and salt-cured beef need to be soaked overnight to attenuate their salty tastes. The beans should be washed and soaked overnight at the same time *(page 16).*

Before being combined with the beans in a stew, all of the salted meats are partially precooked to withdraw remaining excess salt. The meats are then added to

the simmering beans in stages, depending on the time they require to complete their cooking. To preserve its texture, the tongue should be poached by itself and added to the stew at the last minute.

A spicy sauce—prepared by mashing some of the cooked beans with garlic, onions and hot chilies—is added to the stew in its final stages. The sauce acts as a thickener for the stewing liquid and enriches the flavor of the beans and meat.

Perhaps the greatest challenges in presenting a *feijoada* are preparing the side dishes and sauces that accompany the central stew and planning their cooking times so that everything can be served at the same time. Rice—steamed, boiled or made into a basic pilaf as shown on pages 24-25—is an essential side dish. Other traditional accompaniments include hot pepper sauce, marinated onions, sliced oranges, toasted manioc meal and sautéed kale. Their preparations are shown in the box on page 81.

4 Parboiling meats. Place the salt-cured beef in a pot of cold water and bring it to a boil. Simmer it, uncovered, for 30 minutes. Drain and set aside the meat. Place the spareribs, a split pig's foot, and fresh and smoked sausages in another pot of cold water. Bring it to a boil and simmer for 15 minutes. Drain and set aside the meats.

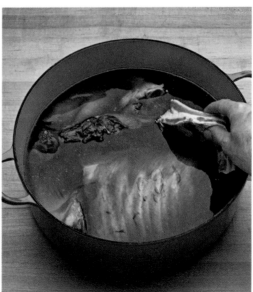

5 Cooking the beans. Drain the beans, pour them into a pot and cover them with water. Boil the beans for 10 minutes, then simmer. Add the salt-cured beef, spareribs, pig's foot and a piece of slab bacon. Simmer for an hour, then add beef chuck and simmer an hour more. Add all of the sausages and simmer for 30 more minutes.

6 Making a sauce. Remove the pot from the heat. Remove 2 cups [½ liter] of beans and 2 cups of liquid and set them aside. Sauté chopped garlic and onions in lard for five minutes. Add peeled, seeded and chopped tomatoes and dried hot chilies, and cook for five more minutes. Then mash in the reserved beans with a wooden spoon.

1 **Soaking the meats.** Place salt-cured beef, smoked beef tongue and salted spareribs in separate bowls of cold water; put the bowls in the refrigerator and let the meats soak overnight. Soak black beans overnight.

2 **Cooking the tongue.** Drain the tongue and place it in a pot of fresh cold water. Bring the water slowly to a simmer, skimming off scum as it surfaces. Add carrots, whole onions, celery ribs and a bouquet garni. With the lid ajar, simmer the tongue for about three hours, or until a fork pierces it easily.

3 **Peeling the tongue.** Drain the tongue. Slit the skin on the underside lengthwise and peel it toward each edge; slice off the skin. On the top, slit the skin crosswise near the base. Peel off the skin toward the tip. Cut away the fat, bone and gristle at the base. Cover the tongue with foil and set it aside.

7 **Simmering the sauce.** Add the reserved bean liquid to the bean sauce. Stirring occasionally, simmer the sauce for about 15 minutes, or until it has the consistency of a porridge *(above)*. Scrape the sauce from its pan into the pot containing the beans.

8 **Refilling the pot.** Mix the bean sauce thoroughly with the contents of the pot. If the liquid level sinks beneath the level of the beans, add boiling water until the contents of the pot are covered. Mix again and simmer gently, stirring occasionally, for about 20 minutes.

9 **Steeping the tongue.** Remove the pot from the heat and skim any fat from the surface of the liquid. Then add the cooked tongue, and allow the meats and beans to steep together off the heat while you finish preparing the other dishes for the meal as shown on the following pages. ▶

10 **Making the pilaf.** Sauté onion slices in olive oil for five minutes; add rice and stir for two or three minutes, until the grains turn a milky color. Stir in peeled, seeded and chopped tomatoes. Add salt and hot chicken stock. Cover the pan and simmer the pilaf for 20 minutes.

11 **Cutting the meats.** To cut the tongue, start at the tip and use a sawing motion to cut the first few slices diagonally. Gradually reduce the angle of each successive cut so that you are cutting directly across the grain where the tongue widens. Cut the remaining meats—except the sausages—into serving pieces.

12 **Presenting the meal.** Arrange all of the meats on a large platter, displaying the sliced tongue in the center, surrounded by the smoked, salted and fresh meats. If you need time to prepare the side dishes shown opposite, cover the meats and keep them in a 200° F. [100° C.] oven. Leave the pilaf and beans covered in their pans. When everything is ready, transfer the beans and rice to separate serving vessels and the accompaniments to small bowls. To serve, arrange all of the bowls and platters attractively, and let the diners help themselves.

A Quintet of Accompaniments

The traditional side dishes for a *feijoada completa* can be prepared in advance and stored until they are needed.

Fiery pepper sauce, for instance, can be made as much as four hours in advance, then covered and refrigerated. Tart, vinegar-dressed onions should be made next; they require three hours of marination. Orange slices stay moist for three hours if kept covered. Toasted manioc meal—made from the bitter cassava plant and sold at Latin American food markets—can be made an hour ahead, covered and left at room temperature. Sautéed kale, like the rice shown on the opposite page, requires 20 to 30 minutes of cooking time; it can be covered and left to wait for half an hour.

Each side dish plays a part in the enjoyment of a *feijoada*. After serving themselves with rice, beans and meat, the diners can add pepper sauce for a hot taste and manioc meal for a bland one. The onions and kale provide tangy notes; oranges refresh the palate.

Pounding pepper sauce. In a mortar, combine finely chopped garlic and onions with dried hot chilies. Use a pestle to pound the ingredients to a coarse paste, then stir in lemon juice. Or, combine all of the ingredients in the bowl of a food processor and whirl until they are coarsely puréed.

Marinating onions. Pour boiling water over sliced onions, drain the onions and rinse them with cold water. Dry the slices. Mix olive oil, vinegar, Tabasco sauce and salt. Stir in the onion slices *(above)* and let them marinate at room temperature for three hours. The onions can be served by themselves or as a garnish for kale *(opposite)*.

Slicing oranges. Use a sharp paring knife to remove the skin and underlying white pith from several oranges. Pull out the core, cut each orange into even slices and discard the end slices. Pick out any seeds. Arrange the slices attractively on a plate.

Cooking manioc meal. Sauté a sliced onion in butter for five minutes. Stir constantly over low heat as you pour in a beaten egg. Gradually stir in manioc meal *(above)*. Cook the meal, stirring often, for eight minutes—or until it is golden. Stir in salt and chopped parsley. Transfer the meal to a bowl and serve it hot or at room temperature.

Sautéing kale greens. Wash and trim kale leaves, removing any bruised spots or tough stems. Shred the leaves and blanch them for three minutes in boiling water. Drain them in a colander, pressing out any excess liquid. Sauté the kale in bacon fat for 20 to 30 minutes, until it is tender. Add salt to taste and place the kale in a serving bowl.

India: An Elaborate Pilaf Imaginatively Garnished

Rice pilaf *(pages 24-25)* is no more than rice that is sautéed, then simmered in liquid. However, elaborations on the dish are legion, and no single country has produced so many or such spectacular versions as India. Indian pilafs, also known as pilaus or *pullaos (recipes, pages 126-129)* may contain such diverse ingredients as fish, chicken, lamb, vegetables, fruits and a range of spices—saffron, cardamom, coriander, turmeric, cinnamon or cloves, for instance. Nonetheless, most such pilafs are characterized by certain basic ingredients and cooking methods.

The rice itself must be long-grain and white; Indians prefer a nutty-flavored type known as basmati, which is sold at specialty food stores. Before using it, spread it out and pick out any visible bits of gravel or other debris. Then wash the rice in several changes of cold water and drain it. This will remove minute debris and rinse starch from the rice, ensuring that the grains will be dry and separate when they are cooked. Brief soaking in cold water softens the rice slightly.

The prepared rice—along with spices, whose flavors are brought out by heat—is traditionally sautéed in *ghee*. This is an Indian version of clarified butter that has a high burning point and nutty flavor. *Ghee* is made by first melting butter over low heat and letting it cook for 45 minutes; the liquid fat is then poured off and the milk solids discarded. Ordinary clarified butter may be used instead.

Vegetables, such as onions, may be added to pilafs when the rice is sautéed—or at any point during the subsequent cooking. Meats, on the other hand, are sautéed separately to give them a brown surface, then added to the pilaf during the last few minutes of cooking. For extra flavor, the meats may be marinated before they are sautéed. The marinade ingredients can be incorporated in the pilaf to enrich it.

Pilafs are served with garnishes that offer counterpoints of taste or texture to the rice and its accompaniments; in India, a pilaf may be presented with as many as a dozen garnishes. Choices include sautéed mushrooms; fried parsley or coriander leaves; broiled tomatoes; small, crisp fried onions; raisins; and shredded fresh coconut.

1 **Trimming the meat.** Put washed long-grain rice in a bowl and add enough cold water to cover it by about 1 inch [2½ cm.]. Let the rice soak for about one hour. Using a sharp knife, trim any excess fat from the meat—here, a boned lamb shoulder—then cut the meat into large cubes.

2 **Coating the meat.** Squeeze the juice from limes; pour it into a shallow dish. Use a mortar and pestle to grind fennel seeds to a coarse powder. Add it to the lime juice. Prick the cubes of meat with a fork and rub them with the lime-and-fennel mixture.

6 **Straining the marinade.** Hold a strainer over the rice, and tip the meat and marinade into the strainer. With the aid of a wooden spoon, push the marinade through the strainer. Set the meat aside. Strain the hot stock into the rice, adding enough stock to cover the rice by about 1¼ inches [3 cm.]. Bring the ingredients to a boil over medium heat.

7 **Adding watercress.** Add finely chopped watercress to the pot. Place a sheet of parchment paper over the top of the pot to trap the steam; cover the pot with its lid. Place the pot in a preheated 350° F. [180° C.] oven and cook the rice for about 12 minutes, until it has absorbed most of the stock.

3 **Adding cream and yogurt.** Whip heavy cream until it just holds its shape. Lightly beat in yogurt. Coat the cubes of meat with the cream-and-yogurt mixture and set them aside.

4 **Preparing spices.** Grind poppy seeds to a fine powder in a mortar and set the powder aside. Bruise whole cloves in the mortar and set them aside, too. Using the blade of a knife, press down firmly on whole cardamoms until the pods split. Then, if you like, splinter one or two cinnamon sticks into large pieces with the knife.

5 **Sautéing the rice.** Drain the rice. Heat clarified butter in a heavy, ovenproof pot until it bubbles; add the poppy seeds, cardamoms, cinnamon and bay leaves. Add the rice and stir over medium heat for two to three minutes, until the rice is translucent. Heat more clarified butter in a pan. Add the cloves, stir in lamb stock, cover and simmer for five minutes. Set the stock aside.

8 **Sautéing the meat.** While the rice cooks, heat clarified butter in a skillet and brown the lamb cubes over medium heat. Remove the rice from the oven and reduce the heat to 325° F. [160° C.]. With a slotted spoon, remove the meat from the skillet and stir it gently into the rice. Replace the parchment paper and lid on the pot and return it to the oven for about 15 minutes.

9 **Garnishing.** Lightly brown blanched, peeled and slivered almonds and blanched, peeled pistachios in clarified butter. Arrange the rice on a serving dish and place the meat on top. Scatter the nuts over the pilaf. Garnish it with raisins that have been soaked in tepid water for an hour, patted dry and sautéed.

Spain: Paella's Marriage of Shellfish and Meats

Of dishes founded on basic rice pilaf *(pages 24-25)*, the paella of Spain ranks among the heartiest: In this rustic assembly, meats, vegetables and seafood are added to the rice in such quantities that the rice assumes an almost subordinate role. Its bulk and mild flavor, however, provide the essential unifying background for the other ingredients.

In Spain, paella is made with a white rice that has a large, plump grain and a slightly sticky texture; American short-grain or medium-grain rice produces the proper effect. As in all pilafs, the rice for paella is heated in fat and then simmered in liquid. But the traditional vessel for cooking a paella—the shallow, two-handled pan from which the dish takes its name—has no lid. Therefore the rice, unlike that in a basic pilaf, is cooked uncovered. Its surface becomes fairly dry—too dry, in fact, for many tastes. If you use a lidless pan, it is advisable to cover the pan with aluminum foil for part of the cooking time.

The paella demonstrated here includes artichokes, tomatoes, green beans and peppers, squid, clams, shrimp, rabbit, smoked sausage and pork. You can substitute peas for beans, mussels for clams, or chicken for rabbit *(recipes, pages 124-125);* you can add other ingredients, such as fish and snails; or you can omit any of the ingredients—except the rice—to make a less elaborate dish.

To ensure that each of the ingredients receives the right amount of cooking, all—except the shrimp, which need only brief cooking—are partly cooked before being combined with the rice *(Steps 1-11)*. When the accompaniments to the rice are almost done, the assembly of the paella begins *(pages 86-87)*. Rice is heated in oil and colored with saffron, then most of the other ingredients are placed on top. Cooking liquid from clams and squid is added, supplemented by water. While the rice simmers in the liquid, the other ingredients finish cooking and their flavors mingle. The ingredients requiring the least amount of extra cooking time—the green beans and clams—are added last and scattered on top of the dish to complete a colorful display.

1 **Preparing the vegetables.** Peel, seed and cut up tomatoes. Roast and peel sweet red peppers—or a mixture of green and red ones—then remove their ribs and seeds, and cut or tear them into strips. Chop onions, garlic cloves and parsley. Trim green beans and cut them into short lengths.

5 **Preparing the squid.** Hold the body pouch of each squid in one hand. With the other hand, grasp the head just below the eyes. Pull the two sections gently apart. Most of the viscera will come away with the head *(above)*. Pull the viscera free and discard them. Reserve the head and body pouch.

6 **Removing the pen.** Run your finger along the rim of the body pouch to locate the tip of its transparent bone, called the pen. Gently pull the pen free and discard it. Rinse the pouch in water to remove the remaining viscera.

7 **Skinning the pouches.** With your fingers, peel the skin off each squid's pouch. From each head, squeeze out the hard beak at the base of the tentacles and discard it. Dip the head in water; press out and discard the eyes.

2 **Trimming the artichokes.** Break off the stem from each artichoke. Pull away the tough, dark outer leaves, snapping them off just above their fleshy bases. To prevent the exposed surfaces of the artichokes from turning black, rub them with half a lemon.

3 **Paring the artichokes.** Cut the top off each artichoke. With a stainless-steel knife, cut spirally from the base to pare away the outer part of the artichoke bottom and to reveal the yellowish flesh. Cut the artichokes into quarters. Cut the fibrous core from each quarter and rub the quarters with lemon.

4 **Cooking the artichokes.** In an enamelled or tin-lined skillet, heat about 1 tablespoon [15 ml.] of olive oil. Gently cook the quartered artichokes for 10 to 15 minutes, turning them from time to time with a wooden spoon. When they have colored, remove them from the heat and set them aside.

8 **Stewing the squid.** Drain the skinned body pouches and the heads. Heat a little butter or oil, sauté the squid briskly for a moment, then reduce the heat and gently stew it until tender—about 20 minutes for squid 10 to 12 inches [25 to 30 cm.] long. Remove the pan from the heat, strain the juices and reserve them separately from the squid.

9 **Cooking the clams.** Scrub live clams thoroughly with a stiff brush. Put them into a saucepan with thyme, parsley, a bay leaf, unpeeled garlic cloves, chopped onion and white wine. Cover and cook over high heat, shaking the pan regularly, until the clams open—five to 10 minutes. Discard any that do not open.

10 **Straining the clams.** Line a colander with cheesecloth or muslin and set it over a large bowl. Tip the contents of the saucepan into the colander and let the clams' cooking liquid drain through. Add to it the squid's stewing liquid. Set the clams aside in their shells. Discard the herbs, garlic and onion. ▶

11 **Preparing pork and tomatoes.** Cut pork Boston shoulder into cubes. Brown the pork in olive oil. Add the previously prepared *(page 84, Step 1)* garlic, onions and tomatoes *(above)*. Cook the mixture briskly, stirring constantly, until most of the liquid has evaporated—about 10 minutes. Set it aside. Prick smoked sausages—preferably *chorizo*—and boil them in water for five minutes to attenuate their smokiness. Drain and slice the sausages and set them aside.

12 **Sautéing the rice.** Cut a rabbit—or a chicken, if you prefer—into eight to 10 pieces. Set aside the liver, if there is one, and brown the cut-up pieces for 10 minutes in olive oil. Add the sausage slices and sauté the meats for another 10 minutes. Set the meats aside. Heat a little olive oil in a wide, heavy skillet, casserole or, as shown, paella pan. Tip a measured quantity of rice—a medium-grain rice is used here—into the pan and then stir in powdered saffron.

15 **Adding shrimp.** Strew the artichoke quarters, pepper slices and squid over the surface of the paella mixture. Arrange raw whole jumbo shrimp on top of the other ingredients, and season the whole mixture with salt and pepper. Set the pan on medium heat and bring the liquid to a boil again. Then cover the pan with aluminum foil and place it in a preheated 350° F. [180° C.] oven to cook for about 10 minutes.

16 **Completing the paella.** Parboil the green bean pieces for two to three minutes and drain them. Remove the paella pan from the oven, uncover it and scatter the beans on top. Stand the shrimp tail-ends up so that the thicker ends of their bodies are buried, and then distribute the cooked clams on top *(above)*. Return the pan to the oven uncovered and cook for about 10 minutes—or until all of the liquid is absorbed.

13 **Adding the meat and tomato to the rice.** With a spatula or wooden spoon, stir the rice gently over medium heat until the grains are coated with oil and have turned a milky color—two to three minutes. Distribute the sausage slices and rabbit or chicken pieces in a layer on top of the rice. Make a second layer with the pork-and-tomato mixture.

14 **Adding the clam liquid.** Dilute the reserved mixture of squid and clam juices with water until the mixture is a little less than twice the volume of the rice in the pan. Do not use double the rice's volume of liquid: Many of the paella ingredients will give off liquid as they cook. Bring the diluted liquid to a boil in a saucepan; pour it over the rice and meat.

17 **Serving.** Thinly slice the reserved rabbit's or chicken's liver, if you have one. Sauté the liver lightly in olive oil until it changes color. Take the paella out of the oven, and garnish it with the liver slices and the previously chopped parsley *(right)*. Serve the paella directly from the pan, making sure that every diner gets some of each ingredient.

America: Jambalaya from the Deep South

Jambalaya is an American Creole variation of a pilaf *(pages 24-25)* — an elaborate rice-based dish containing a diversity of ingredients. Although probably named for *jamón* — the Spanish word for ham — a jambalaya typically includes an assortment of meats, seafood and vegetables, chosen according to availability and personal tastes *(recipe, page 125)*. In Louisiana, cooks exploit the abundant local shellfish — most notably crabs, crayfish and shrimp — and sometimes add chicken and sausage as well.

The basic techniques for a jambalaya parallel those for any rice pilaf, but the Creole tradition calls for a higher ratio of liquid to rice — as much as three to one — which results in a moister consistency than usual. You can, of course, overrule tradition and reduce the liquid-to-rice ratio if you prefer a drier consistency.

The jambalaya demonstrated here includes blue crabs in their shells and shucked oysters. The live crabs must be rinsed, dispatched with the point of a sharp knife and carefully cleaned before the assembly begins. The oysters, added at the last minute after everything else has cooked, simply steep in the hot stew for a few minutes before serving.

1 Preparing the crabs. Rinse the crabs and puncture each one deeply behind the eyes with a sharp knife. To prepare each crab, first remove the claws and turn the crab upside down to pry off the triangular flap on its underside *(above, left)*. Then gently separate the top shell from the lower half of the crab *(center)*. Remove the gills and any spongy material from the exposed top of the crab's body *(right)* and from the underside of the top shell. Crack the claws and set the cleaned crabs aside in a bowl.

2 Combining ingredients. Sauté chopped onions and green pepper in butter for five minutes. Add rice and chopped ham and stir until the rice turns a milky color. Add the crabs and stir. Add peeled, seeded and chopped tomatoes, a bay leaf and fresh thyme. Chop fresh parsley and set it aside.

3 Adding hot water. Pour in hot water, using two or three times the volume of rice, depending on the consistency desired. Stir once, bring to a boil and cover. Simmer over low heat for 20 to 30 minutes.

4 Steeping oysters. Add chopped parsley for the last five minutes of cooking. Then turn the heat off and add shucked raw oysters with their liquor *(above)*. Cover the pan and let the oysters steep for two or three minutes.

5 Serving. Transfer the stew to a bowl; season with Tabasco sauce and arrange the crabs around the edge. Each serving should include some of the rice mixture, oysters and a crab.

Anthology of Recipes

Drawing upon the cooking traditions and literature of more than 40 countries, the editors and consultants for this volume have assembled 200 published recipes for the Anthology that follows. The selections range from the familiar to the exotic—from nine varieties of so basic a dish as pilaf to a novel recipe for a cake that substitutes ground navy beans for flour.

Many of the recipes were written by world-renowned exponents of the culinary art, but the Anthology also includes selections from rare and out-of-print books and from works that have never been published in English. Whatever the sources, the emphasis in these recipes is always on natural ingredients.

Since many early recipe writers did not specify amounts of ingredients, sizes of pans, or even cooking times and temperatures, the missing information has been judiciously added. In some cases, instructions have been expanded, but where the directions still seem abrupt, the reader need only refer to the demonstrations in the front of the book. Where appropriate, clarifying notes have also been supplied; they are printed in italics. Modern terms have been substituted for archaic language; but to preserve the character of the original recipes, the authors' texts have been changed as little as possible.

The amount of cooking a particular batch of beans, lentils, peas or grain will require depends on factors that may be beyond control: variety, growing conditions, drying method and length of storage. Cooking times should thus be considered only as guidelines and the dish repeatedly tested for doneness.

Because most beans and many grains are interchangeable, the recipes are categorized by families. For easy reference, bean-curd recipes are grouped together. Recipes for standard preparations—tomato sauce and stocks among them—appear at the end of the Anthology. Unfamiliar cooking terms and ingredients are explained in the General Index and Glossary.

Apart from the primary components—beans, bean curd, lentils, peas or grains—all ingredients are listed within each recipe in order of use, with both U.S. and metric measurements provided. The metric quantities supplied here reflect the American practice of measuring such solid ingredients as flour or sugar by volume rather than, as European cooks do, by weight.

To make the quantities simpler to measure, many of the figures have been rounded off to correspond to the gradations that are now standard on U.S. metric spoons and cups. (One cup, for example, equals 237 milliliters; however, where practicable in these recipes, a cup's equivalent appears as a more readily measured 250 milliliters—¼ liter.) Thus the American and metric figures do not precisely match, but using one set or the other will produce the same good results.

Beans, lentils and peas need boiling to destroy toxins (box, page 15).

Beans

Ginger-flavored Chinese-Style Adzuki Beans

To serve 4

2 cups	dried adzuki beans	½ liter
1	garlic clove	1
5 or 6	peppercorns	5 or 6
1-inch	piece fresh ginger	2½-cm.
1-inch	cinnamon stick	2½-cm.
10	coriander seeds	10
2	whole cloves	2
	star anise (optional)	
1 tsp.	salt	5 ml.
1	onion	1
¼ cup	molasses or plum jelly	50 ml.
1 to 2 tbsp.	sweet sherry	15 to 30 ml.
1 tbsp.	*tamari* soy sauce	15 ml.

Put half of the adzuki beans into a pot with 3 cups [¾ liter] of water, and bring them to a boil. Tie the garlic, peppercorns, ginger, cinnamon, coriander, cloves and a small piece of the star anise, if using, into a spice bag. Put the bag into the bean pot along with the salt, the remaining adzuki beans and the onion. Bring the mixture to a boil again, then cover and cook for about two hours, or until the beans are tender. Remove the onion and the spice bag. Stir in the molasses or jelly, sherry and *tamari* soy sauce.

SUSAN RESTINO
MRS. RESTINO'S COUNTRY KITCHEN

Brazilian Black Bean Stew
Feijoada Completa

The techniques of making this stew and assembling the various side dishes are shown on pages 78-81. As a substitute for corned spareribs, you may pack fresh spareribs in coarse salt, leave them overnight in the refrigerator, then wipe them free of salt when you are ready to use them. The salt-cured beef specified in this recipe is seasoned, sun-dried meat, sold as tasajo in Latin American food markets. Manioc meal, ground from the root of the cassava plant, is also available in the same markets. The volatile oils in hot chilies may make your skin sting and your eyes burn; after handling chilies, avoid touching your face and wash your hands promptly.

To serve 8 to 10

4 cups	dried black beans, soaked overnight and drained	1 liter
3 lb.	smoked beef tongue, soaked overnight and drained	1½ kg.
1 lb.	corned spareribs, soaked overnight and drained	½ kg.
½ lb.	salt-cured beef, soaked overnight and drained	¼ kg.
1	celery rib	1
2	carrots	2
4	medium-sized onions, 2 coarsely chopped	4
1	bouquet garni of parsley, leek, bay leaf and tarragon	1
½ lb.	*chorizo*	¼ kg.
½ lb.	fresh pork sausage links	¼ kg.
1	pig's foot, split	1
¼ lb.	slab bacon, rind removed	125 g.
1 lb.	lean boneless beef chuck	½ kg.
2 tbsp.	lard	30 ml.
1 tbsp.	finely chopped garlic	15 ml.
3	medium-sized tomatoes, peeled, seeded and coarsely chopped	3
2	dried hot chilies	2
1 tsp.	salt	5 ml.
½ tsp.	freshly ground pepper	2 ml.
2	oranges, peeled and sliced	2

Toasted manioc meal with egg

2 tbsp.	butter	30 ml.
½	large onion, thinly sliced	½
1	egg, lightly beaten	1
1⅓ cups	manioc meal	325 ml.
1 tsp.	salt	5 ml.
1 tbsp.	finely chopped fresh parsley	15 ml.
4	pimiento-stuffed olives, sliced (optional)	4
2 to 4	eggs, hard-boiled and halved lengthwise (optional)	2 to 4

Shredded kale

1	large onion, sliced into rings	1
3 tbsp.	olive oil	45 ml.
3 tbsp.	red wine vinegar	45 ml.
3 tbsp.	Tabasco sauce	45 ml.
1¾ tsp.	salt	9 ml.
5 lb.	kale or collard greens	2½ kg.
¾ cup	rendered bacon fat	175 ml.

Pepper-and-lemon sauce

3 or 4	dried hot chilies	3 or 4
½ tsp.	salt	2 ml.
1	onion, sliced	1
1	garlic clove	1
½ cup	strained fresh lemon juice	125 ml.

Rice with tomatoes and onion

¼ cup	olive oil	50 ml.
1	large onion, thinly sliced	1
3 cups	unprocessed long-grain white rice	¾ liter
3 cups	chicken stock (recipe, page 166)	¾ liter
2	medium-sized tomatoes, peeled, seeded and coarsely chopped	2
1 tsp.	salt	5 ml.

Cover the tongue with fresh water, add the celery, carrots, two whole onions and bouquet garni, and bring the water to a boil over high heat; skim off any scum, partially cover the pan, reduce the heat to low, and—periodically adding more water if necessary to keep the tongue covered—simmer the tongue for three to four hours, until it is tender. Discard the aromatics and skin the tongue with a knife, cutting away all the fat, bones and gristle at its base.

Cover the salt-cured beef with water and bring to a boil. Reduce the heat and simmer, uncovered, for 30 minutes. Cover the spareribs, *chorizo*, fresh sausage and pig's foot with water. Bring to a boil, reduce the heat and simmer, uncovered, for 15 minutes. Drain the meats and set aside.

In a large pot, combine the beans, salt-cured beef, spareribs, pig's foot and bacon. Cover with 3 quarts [3 liters] of water and bring to a boil. Reduce the heat, cover and simmer for one hour. The water should cook away somewhat, leaving the beans moist and slightly soupy. If the beans are getting too dry, add boiling water.

Add the beef chuck to the pot and cook the beans and meat for one hour. Finally, add the *chorizo* and fresh sausage and cook for 30 minutes. Skim the fat from the surface of the beans, add the tongue and remove the pot from the heat.

In a heavy skillet, melt the lard and cook the chopped onions and garlic over moderate heat, stirring frequently, for five minutes, until the onions are soft and transparent but not brown. Stir in the tomatoes, dried chilies, salt and pepper, and simmer for five minutes. With a slotted spoon, remove 2 cups [½ liter] of the beans from the casserole and add them to the skillet. Mash them thoroughly into the onion mixture, moistening them with 2 cups of the bean cooking liquid as you mash. Stirring occasionally, simmer this sauce over low heat for 15 minutes, or until it thickens. With a rubber spatula, scrape the sauce into the bean pot, and cook over low heat, stirring occasionally, for 20 minutes.

For the toasted manioc meal with egg, melt the butter over medium heat in a heavy skillet, tipping the pan to coat the bottom evenly. Cook the onion slices, stirring constantly, until they are soft. Reduce the heat to low and—still stirring constantly—pour in the egg. The egg will coagulate in seconds. Slowly stir in the manioc meal and cook, stirring frequently, for eight minutes, or until the meal becomes golden. Stir in the salt and parsley. Serve hot or at room temperature, topped with either the olives or hard-boiled eggs.

For the shredded kale, first parboil the onion rings for one minute, drain them, and marinate them for three hours in the olive oil, vinegar, Tabasco sauce and ¼ teaspoon [1 ml.] of the salt. Wash the greens under cold running water. Cut the leaves from their tough stems and shred them into ¼-inch [6-mm.] strips. Cook the greens in a large pot of boiling water for three minutes and drain them.

In a heavy skillet, melt the bacon fat and, when it is hot, add the greens. Cook, stirring frequently, for 30 minutes, or until the greens are tender but slightly crisp. Do not let them brown. Stir the remaining 1½ teaspoons [7 ml.] of salt into the greens and serve them at once with a garnish of the marinated onions. If the greens must wait, cover them with foil and keep them warm in a preheated 200° F. [100° C.] oven. Garnish them at the last minute.

For the pepper-and-lemon sauce, pound the chilies, salt, onion and garlic to a paste; add the lemon juice, and let the sauce stand for one hour at room temperature or refrigerate it, covered, for as many as four hours. The sauce must then be used: It will ferment if it is stored longer.

For the rice with tomatoes and onion, heat the oil in a saucepan. Cook the onion until it is translucent. Pour in the rice and stir for two to three minutes, until all the grains are coated with oil. Stir in the stock, 3 cups [¾ liter] of boiling water, tomatoes and salt, and bring to a boil. Cover the pan and reduce the heat to very low. Simmer for 20 minutes, or until the rice has absorbed all the liquid. If the rice must wait, drape the pan with a towel and keep it warm in a preheated 200° F. [100° C.] oven.

To present the *feijoada*, remove the meats from the pot. Slice the tongue, dried beef, bacon, chuck, spareribs and pig's foot into serving pieces and separate the two sausages. Transfer the beans to a serving bowl. Present all the meats on one large, heated platter with the sliced tongue in the center, the fresh meat on one side, the smoked meats on the other. Present the beans, the orange slices and the side dishes of toasted manioc meal, shredded kale, pepper-and-lemon sauce, and rice in separate bowls or platters.

Beans, lentils and peas need boiling to destroy toxins (box, page 15).

Well-fried Beans with Totopos

Frijoles Refritos

The technique of making frijoles refritos is demonstrated on pages 44-45. The totopos used to garnish the beans are triangular pieces of corn tortilla fried in lard until golden brown. Refried beans can also be served with chilaquiles, strips of corn tortilla that are fried, then baked in chili sauce.

To cut a radish into a rose for the garnish, make two slits in the shape of a cross in the top of the radish, then another pair of slits diagonally through the cross to form a star. Soak the radish in ice water for 15 minutes; the flesh will expand to form a rose.

To serve 6

2 cups	dried black, pink or pinto beans	½ liter
about ⅓ cup	finely chopped onion	about 75 ml.
8 tbsp.	lard or pork drippings	120 ml.
	salt	
¼ cup	farmer or pot cheese, broken up with a fork	50 ml.
12	totopos	12
	romaine lettuce leaves	
6	radishes, cut into roses	6

Put the beans into a deep pan and cover them with about 6 cups [1½ liters] of cold water. Add half of the chopped onion and 2 tablespoons [30 ml.] of the lard or drippings, and bring them to a boil. Reduce the heat, cover the pan, and barely simmer the beans, without stirring, until they are tender but not soft—about two hours for black beans, one and one half hours for other varieties.

Add a large pinch of salt and simmer the beans for another 30 minutes. Set them aside, preferably until the next day. There should be about 4 cups [1 liter] of cooking liquid.

Heat the remaining lard or drippings, and fry the rest of the onion until it is soft but not browned. Add 1 cup [¼ liter] of beans in their cooking liquid to the pan, and mash them well over very high heat. Add the rest of the beans gradually, mashing them all the time, until you have a coarse purée. When the purée begins to dry out and sizzle at the edges, it will start to come away from the surface of the pan. As you let it continue cooking, tip the pan from side to side. The purée will form itself into a loose roll. This process will take from 15 to 20 minutes.

Tip the roll, rather as if folding an omelet, onto a serving dish, garnish with the cheese, and spike it with the *totopos*. Garnish the roll with the lettuce and radishes, and serve the dish immediately.

DIANA KENNEDY
THE CUISINES OF MEXICO

Black Bean Soufflé

The cooking of black beans is explained on pages 16-17.

To serve 4

1½ cups	cooked black beans	375 ml.
1 tbsp.	grated orange peel	15 ml.
1 tbsp.	dried mint leaves	15 ml.
	salt (optional)	
5	egg whites, stiffly beaten	5
1 cup	sour cream	¼ liter
2 tbsp.	orange-flavored liqueur	30 ml.

Purée the beans in a food mill with the orange rind, mint and a little salt if you like.

Put one large spoonful of the egg whites into the black bean purée and stir well. Fold the rest of the egg whites in gently. Turn the mixture into a buttered 5-cup [1¼-liter] soufflé dish, and bake in a preheated 350° F. [180° C.] oven for 35 to 45 minutes. Serve the soufflé immediately with a sauce made by thinning and flavoring the sour cream with the orange liqueur.

MARIAN TRACY
REAL FOOD

Baked Black Beans

To serve 6

2 cups	dried black beans, soaked overnight	½ liter
1	large onion, sliced	1
1	garlic clove	1
2	celery ribs	2
1	bay leaf	1
	fresh parsley sprigs	
3 tbsp.	finely chopped salt pork	45 ml.
	salt and pepper	
	cayenne pepper	
¼ cup	dark rum	50 ml.
	meat stock (recipe, page 166) (optional)	
	sour cream	

Add to the beans the onion, garlic, celery, bay leaf and several sprigs of parsley. Cook the beans until they are almost tender, about 1½ hours, adding more water if necessary to keep the beans covered. Remove the garlic, celery and herbs.

In a small skillet, fry the salt pork until the scraps are golden and add them, with the melted fat, to the beans. Then add salt, pepper, a pinch or more of cayenne to taste and the rum. Transfer the beans to a casserole and, if there is not enough liquid to almost cover them, add a little more water or some meat stock.

Bake the beans, covered, in a preheated 300° F. [150° C.] oven for one hour, stirring occasionally, or until they are very tender—but not mushy—and the juice is reduced. Serve the beans with a large sauceboat of cold sour cream.

NARCISSE CHAMBERLAIN AND NARCISSA G. CHAMBERLAIN
THE CHAMBERLAIN SAMPLER OF AMERICAN COOKING

Bean and Sauerkraut Casserole

Babos Káposztafözelék

To serve 6

1½ cups	dried cranberry beans, soaked overnight and drained	375 ml.
1 lb.	smoked pork ribs	½ kg.
2 cups	sauerkraut, drained	½ liter
2 cups	puréed tomato *(recipe, page 166)*	½ liter
1	large onion, coarsely chopped	1
2 tbsp.	lard	30 ml.
2 tbsp.	flour	30 ml.

Place the beans and pork ribs in a large pot, cover them with water, bring to a boil and skim. Cover the pot, and cook the beans for one and one half to two hours until they are tender.

Meanwhile, simmer the sauerkraut and tomato together for about 10 minutes. In a saucepan, sauté the onion in the lard over medium heat until soft; sprinkle with the flour and cook, stirring constantly, until light golden in color. Add the onion roux to the sauerkraut, mix well, cover, and simmer over medium heat for about 20 minutes until the juices have slightly thickened.

Drain the beans and mix them, plus the pork ribs, with the sauerkraut.

GEORGE LANG
THE CUISINE OF HUNGARY

Flageolets

The author suggests that these beans be served alongside roast leg of lamb.

To serve 4

1 cup	dried flageolets, soaked overnight in 2 cups [½ liter] water	¼ liter
1 tsp.	sea salt	5 ml.
2 tbsp.	olive oil	30 ml.
½ tsp.	finely chopped garlic	2 ml.
2 or 3	medium-sized tomatoes, peeled and quartered	2 or 3
	herb salt	

Put the beans in a pan with the soaking water and sea salt. Bring the beans to a boil, reduce the heat, cover, and allow them to simmer until they are soft—one and one half to two hours. Drain the beans. Heat the olive oil with the chopped garlic in a deep, heavy pan, and cook over low heat for two or three minutes. Add the tomatoes and the cooked flageolets, and season them with some herb salt. Cover, and cook over low heat for 20 minutes.

MARION GORMAN AND FELIPE P. DE ALBA
THE DIONE LUCAS BOOK OF NATURAL FRENCH COOKING

Flageolets Breton-Style

Flageolets à la Bretonne

Dried navy beans may also be prepared in this manner.

To serve 6

2 cups	dried flageolets, soaked overnight and drained	½ liter
1	large onion, finely chopped	1
2 tbsp.	butter	30 ml.
⅓ cup	dry white wine	75 ml.
3	medium-sized tomatoes, peeled, seeded and chopped	3
	salt and pepper	
¼ cup	chopped fresh parsley	50 ml.

Place the beans in a large pot, cover them with water, cover the pot, and cook the beans for one and one half to two hours, or until tender. Drain the beans.

Cook the onion in the butter until golden. Add the wine and chopped tomatoes. Bring the mixture to a boil, reduce the heat to low and cook for 10 to 15 minutes, or until the tomatoes are soft and beginning to disintegrate. Add the beans, season with salt and pepper, and stir in about half of the parsley. Cover, and simmer for 20 minutes. Serve garnished with the remaining parsley.

LE CORDON BLEU

Beans, lentils and peas need boiling to destroy toxins (box, page 15).

Fava Bean Salad
Fool Imdamis

To serve 4

2 cups	dried fava beans, soaked overnight, drained and peeled	½ liter
1	onion, chopped	1
1	garlic clove, chopped	1
2	tomatoes, quartered	2
1 tsp.	dried mint	5 ml.
2 tbsp.	lemon juice	30 ml.

Cover the beans with water by 2 inches [5 cm.], bring to a boil, and cook them for one and one half to two hours until tender. Drain, cool the beans, then add the onion, garlic, tomatoes, mint and lemon juice. Toss lightly and serve.

HELEN COREY
THE ART OF SYRIAN COOKERY

Golden Gram Pancakes
Bindae Duk

Kimchi, a spicy Korean condiment of pickled cabbage, is available at Korean markets and some supermarkets.

To make about 20 pancakes

1 cup	dried golden gram, soaked overnight and drained	¼ liter
2	eggs, beaten	2
¼ lb.	lean ground pork	125 g.
1	small onion, finely chopped	1
1	scallion, finely chopped	1
2	garlic cloves, crushed	2
1 tsp.	salt	5 ml.
¼ tsp.	pepper	1 ml.
1 tsp.	peeled and grated fresh ginger	5 ml.
½ cup	chopped bean sprouts	125 ml.
½ cup	chopped *kimchi* or Chinese cabbage	125 ml.
2 tbsp.	Chinese sesame-seed oil	30 ml.
2 tbsp.	oil	30 ml.

Purée the gram with 1 cup [¼ liter] of water in a blender until smooth. Pour the purée into a bowl, add the other ingredients and mix well. Oil a griddle or heavy skillet, heat it, and drop tablespoonfuls of the mixture onto the hot surface. Cook the pancakes until they are golden brown underneath; turn them and cook the other side. Serve them hot or cold.

CHARMAINE SOLOMON
THE COMPLETE ASIAN COOKBOOK

Vegetables and Golden Gram

The volatile oils in hot chilies may make your skin sting and your eyes burn; after handling chilies, avoid touching your face and wash your hands promptly.

To serve 6

2½ cups	dried golden gram	625 ml.
2 tsp.	turmeric	10 ml.
1 tsp.	salt	5 ml.
4 tbsp.	*ghee* or oil	60 ml.
4	potatoes, cut into large pieces	4
2	carrots, cut into large pieces	2
¼ lb.	cauliflower, separated into florets	125 g.
1 tsp.	sugar	5 ml.
3 to 6	fresh green beans, cut into 1½-inch [4-cm.] pieces	3 to 6
1 cup	chopped cabbage leaves	¼ liter
2	radishes, chopped	2
3	tomatoes, halved	3
2 tbsp.	shelled fresh peas	30 ml.
3 or 4	fresh hot green chilies, sliced	3 or 4
2 tsp.	chopped fresh coriander leaves	10 ml.
4	dried hot chilies	4
½ tsp.	ground cumin	2 ml.
two 2-inch	cinnamon sticks	two 5-cm.
4	cardamom pods	4
3	bay leaves	3
1 tbsp.	peeled and grated fresh ginger	15 ml.
4	onions, sliced	4
1 tsp.	*garam masala*	5 ml.

Lightly brown the gram in a clean, dry pan over low heat. Rinse the beans, then bring them to a boil in 4 cups [1 liter] of water, with the turmeric and salt, in a covered pan. Reduce the heat so that the beans simmer.

While the beans cook, heat 1 tablespoon [15 ml.] of the *ghee* or oil in a large skillet, and in it stir fry the potatoes, carrots and cauliflower until the potatoes have turned light brown. Remove this vegetable mixture to a dish.

When the gram is three quarters cooked (after about 30 minutes), add to it the sugar, green beans, cabbage and radishes. When these vegetables are tender (after about 15 minutes), add the tomatoes, peas, green chilies and coriander leaves, and remove the bean mixture from the heat.

Heat another 1 tablespoon of the *ghee* or oil in the skillet over medium heat, and fry the dried chilies, cumin, cinnamon and cardamom for one minute. Add the bay leaves, ginger and onions, and stir fry this seasoning mixture for about 10 minutes. Combine the vegetables, beans and seasonings in the skillet, and simmer everything together for 10 minutes. Finally, add the *garam masala,* remove from the heat and serve, if you like, with rice.

P. MAJUMDER
COOK INDIAN

Spinach and Golden-Gram Dumplings

The technique of making these dumplings is demonstrated on pages 54-55. The volatile oils in hot chilies may make your skin sting and your eyes burn; after handling chilies, avoid touching your face and wash your hands promptly.

These golden-gram dumplings, streaked with shreds of spinach leaves, are a delicacy from the North Indian state of Uttar Pradesh. They can be made several hours ahead and, just before serving, they can be reheated, loosely covered, in the middle of a preheated 350° F. [180° C.] oven for 12 to 15 minutes.

To serve 6		
1 cup	dried golden gram, soaked for 4 hours, rinsed and drained	¼ liter
½ cup	shredded spinach, firmly packed	125 ml.
1 tbsp.	finely chopped coriander leaves	15 ml.
1 or 2	fresh hot green chilies, stemmed, seeded and thinly sliced	1 or 2
⅛ tsp.	baking powder	½ ml.
¾ tsp.	coarse salt	4 ml.
	peanut or corn oil for deep frying	

Mint-coriander sauce		
¼ cup	fresh mint leaves, packed	50 ml.
¾ cup	fresh coriander leaves, packed	175 ml.
⅓ cup	yogurt	75 ml.
1 tbsp.	finely chopped onions	15 ml.
⅓ tsp.	peeled and grated fresh ginger	1½ ml.
1 or 2	fresh hot green chilies, stemmed and seeded	1 or 2
about 1 tsp.	coarse salt	about 5 ml.
¾ tsp.	sugar	4 ml.
¼	green pepper, seeded and deribbed	¼

To prepare the mint-coriander sauce, put all of the sauce ingredients into the container of a food processor or electric blender, and blend them until they are finely puréed and reduced to a creamy sauce. Taste the sauce for salt, and pour it into a small bowl. Cover the bowl, and place the sauce in the refrigerator to chill thoroughly.

To make the dumplings, put the gram in a food processor with ½ cup [125 ml.] of water and process it for five to six minutes, turning the machine on and off every 15 to 20 seconds and scraping down the sides of the container from time to time. Remove the resulting paste to a mixing bowl and stir in the spinach, coriander leaves, chilies, baking powder and salt. Do not overblend or the mixture will become dense, which will in turn make the dumplings hard and chewy.

Heat the oil in a deep fryer until it is very hot but not smoking—375° F. [190° C.]. Drop the gram mixture, a heaping teaspoonful at a time, into the hot oil. Fry eight to 12 dumplings at a time—making certain not to overcrowd them—until they are light golden—about four to five minutes. Take them out with a slotted spoon and drain them on paper towels. Serve the dumplings hot, accompanied by the chilled sauce.

JULIE SAHNI
CLASSIC INDIAN COOKING

Meat Casserole with White Beans

Mon Favori

To serve 4		
1 cup	dried Great Northern or navy beans, soaked overnight and drained	¼ liter
2 tbsp.	butter	30 ml.
10 oz.	lean pork, cut into pieces	300 g.
10 oz.	veal, cut into pieces	300 g.
½ lb.	turkey or chicken gizzards, hearts, necks and wing tips	¼ kg.
4	small smoked sausages	4
2	tomatoes, halved, squeezed, juice strained and reserved	2
	salt	

Heat the butter in a heavy pot. Add the pork and sauté lightly, then add the veal, giblets and sausages; sauté until all the meat is golden.

Meanwhile, cook the beans for 30 minutes in enough water to cover them. Add the partially cooked beans, together with their cooking liquid, to the pot containing the meats—there should be enough liquid to cover all of the ingredients. If there is not, add more water. Finally, stir in the tomato juice. Cover the pot and cook the mixture over low heat for three hours. Add salt to taste before serving.

ÉDOUARD M. NIGNON (EDITOR)
LE LIVRE DE CUISINE DE L'OUEST-ÉCLAIR

Beans, lentils and peas need boiling to destroy toxins (box, page 15).

White Bean Salad
Salade de Haricots Blancs

To serve 8

1 lb.	dried Great Northern or navy beans, soaked overnight and drained	½ kg.
2	onions, quartered	2
2	celery ribs	2
5 tbsp.	strained fresh lemon juice	75 ml.
1 tsp.	salt	5 ml.
	freshly ground pepper	
3 tbsp.	chopped fresh parsley	45 ml.
3 tbsp.	capers	45 ml.
1½ tbsp.	chopped fresh thyme or rosemary leaves	22 ml.
¾ cup	olive oil	175 ml.

Place the beans in a large pot, cover with water and bring it to a boil. Add the onions and celery ribs. Cover the pot, and cook the beans for one and one half to two hours, or until they are tender. Discard the onions and celery and drain the beans. Beat together the remaining ingredients to make a dressing, and pour it over the beans while they are still hot. Toss carefully, allow the beans to cool, and serve the salad at room temperature.

MARY HENDERSON
PARIS EMBASSY COOKBOOK

White Beans in Red Wine Sauce
Haricots Blancs au Vin Rouge

To serve 4

2 cups	dried Great Northern or navy beans, soaked for 2 hours in warm water	½ liter
	salt	
1	onion, stuck with 1 whole clove	1
1	small carrot	1
1	bouquet garni	1
3	shallots, chopped	3
5 tbsp.	butter	75 ml.
¾ cup	dry red wine	175 ml.
1	garlic clove, chopped	1
	freshly ground pepper	
2 tbsp.	chopped fresh parsley	30 ml.

Bring the beans to a boil in their soaking water and skim the surface. Add the onion, carrot and bouquet garni, and sim-

mer the beans about one hour, or until tender. In a heavy casserole, sauté the shallots with half of the butter, covered, for two minutes over low heat. Add the red wine, bring to a boil, and cook until the bottom of the casserole is nearly dry.

Drain almost all of the liquid from the beans, reserving a few spoonfuls. Add the beans to the casserole. Stir in the garlic and cook over low heat for 10 minutes. Add pepper and taste for seasoning.

Cut the remaining butter into pieces the size of hazelnuts and stir them, one by one, into the beans. Gently sauté the beans until the butter is distributed evenly. The beans should bathe lightly in a broth; if they do not, add a few spoonfuls of the reserved cooking liquid. Sprinkle with the parsley and serve.

JEAN AND PIERRE TROISGROS
THE NOUVELLE CUISINE OF JEAN & PIERRE TROISGROS

Greek Baked Beans
Fasolia Plaki

Kefalotiri is a hard goat's-milk cheese similar to Parmesan. It is sold at Greek food markets.

To serve 4 to 6

1½ cups	dried Great Northern, navy or other beans, soaked overnight and drained	375 ml.
¼ cup	oil	50 ml.
2	onions, chopped	2
1	garlic clove, chopped (optional)	1
1 cup	puréed tomato (recipe, page 166)	¼ liter
2 tbsp.	vinegar	30 ml.
3 tbsp.	honey	45 ml.
1	bay leaf	1
2	whole cloves	2
¼ cup	grated *kefalotiri* or Parmesan cheese	50 ml.

Place the beans in a large pot, cover with water, cover the pot, and cook the beans for one and one half to two hours, or until they are tender. Drain the beans.

Heat the oil in a saucepan and fry the onions until they are soft. Add all of the remaining ingredients except the cheese, plus 2 cups [½ liter] of water, and bring them to a boil to make a sauce.

When the beans are cooked, drain them. Mix the sauce with the beans and pour the mixture into a shallow baking pan. Bake for 30 minutes in a preheated 375° F. [190° C.] oven. Sprinkle the top with cheese and serve.

THERESA KARAS YIANILOS
THE COMPLETE GREEK COOKBOOK

Mixed-Meat and Bean Stew
Caldo Gallego

To serve 6 to 8

1 cup	dried Great Northern or navy beans, soaked overnight and drained	¼ liter
½ lb.	smoked ham	¼ kg.
½ lb.	boiled or baked ham	¼ kg.
1 lb.	veal	½ kg.
¼ lb.	slab bacon or salt pork	125 g.
½ lb.	pork sausage	¼ kg.
¼	stewing chicken (optional)	¼
	salt and pepper	
½	medium-sized cabbage, coarsely chopped	½
4	turnips, each cut in 2 pieces	4
about ½ cup	turnip greens, firmly packed	about 125 ml.
1 lb.	potatoes, quartered	½ kg.
½	medium-sized onion, sliced	½
1	*chorizo* or pepperoni (about ⅓ lb. [175 g.])	1

Put the hams, veal, bacon, pork sausage and chicken in an earthenware or other large, heavy pot; cover the meats with cold water, bring to a boil, and skim. Season the mixture with salt and pepper, and simmer, covered, as slowly as possible for four hours.

One hour after putting on the meat, place the beans in a second pot, pour in cold water to the level of the beans, and add 4 cups [1 liter] more water. Simmer the beans slowly until they are almost tender—about one and one half to two hours—then add the cabbage, turnips, turnip greens, potatoes, onion, and the whole *chorizo* or pepperoni. Continue simmering slowly, adding boiling water if necessary to keep the food covered.

When the vegetables are almost cooked, combine the vegetables and their broth with the meat; season the mixture, and let it all simmer together another half hour.

At serving time, ladle off the liquid and serve it first as a soup; follow the soup with the veal, ham, pork sausage, bacon, and chicken cut into serving pieces on one platter and the vegetables and beans on a second platter, with the *chorizo* sliced on top.

BARBARA NORMAN
THE SPANISH COOKBOOK

White Bean and Lamb Stew
Haricot de Mouton

To serve 6

3 cups	dried Great Northern or navy beans, soaked overnight and drained	¾ liter
4 tbsp.	butter	60 ml.
1 lb. each	boned lamb neck, breast and shoulder, cut into pieces	½ kg. each
2	turnips, thinly sliced	2
4	garlic cloves, crushed	4
¼ cup	flour	50 ml.
	salt and freshly ground pepper	
1	bouquet garni	1
3	tomatoes, peeled, seeded and coarsely chopped	3
½ lb.	salt pork, cut into lardons, parboiled in boiling water for 1 minute, dried and sautéed in butter	¼ kg.
½ lb.	small boiling onions, sautéed in butter until golden	¼ kg.

Place the beans in a large pot, cover them with water, cover the pot, and cook the beans for one and one half to two hours, or until they are tender. Drain the beans.

Heat the butter in a pan, add the meat and turnips, and cook over medium heat until the turnips are golden and the meat is well browned. Add the garlic and the flour, and cook for five minutes. Pour in about 4 cups [1 liter] of water, season with salt and pepper, and add the bouquet garni and the tomatoes. Cover the pan and simmer over low heat for 50 minutes.

Pour the contents of the pan into a large sieve placed over a bowl. Transfer the pieces of meat to a casserole, scatter the bacon lardons and onions over the meat, and spoon the beans on top. Degrease the strained meat juices and pour them over the contents of the casserole. Bring the liquid to a boil, then cover the casserole and cook in a preheated 325° F. [160° C.] oven for one hour.

ÉDOUARD NIGNON (EDITOR)
LE LIVRE DE CUISINE DE L'OUEST-ÉCLAIR

Beans, lentils and peas need boiling to destroy toxins (box, page 15).

Cassoulet

The technique of making this dish is demonstrated on pages 74-77. The author suggests that the best garlic sausage for this dish is cervelas—a large, French poaching sausage—or Polish kielbasa. They are available at specialty meat markets.

	To serve 6 to 8	
4 cups	dried Great Northern beans, soaked overnight and drained	1 liter
	mixed herbs (savory, thyme, oregano, rosemary and bay leaf)	
2 lb.	goose (breast or leg)	1 kg.
	salt	
	goose fat from inside the bird	
6 oz.	fresh pork rind, rolled and tied with a string	175 g.
½ lb.	lean salt pork	¼ kg.
1	pig's foot	1
2	carrots, cut into pieces	2
1	large onion stuck with 2 whole cloves	1
2	garlic cloves	2
1	bouquet garni	1
1	garlic poaching sausage, pricked in several places with a fork	1
	salt	

Lamb stew

2	medium-sized carrots, cut into ¾-inch [2-cm.] pieces	2
2	medium-sized onions, coarsely chopped	2
	salt	
3 lb.	lamb shoulder, surface fat trimmed off, meat cut into large pieces but not boned	1½ kg.
2 tbsp.	flour	30 ml.
1 cup	dry white wine	¼ liter
5	garlic cloves	5
1	bouquet garni	1
3 or 4	tomatoes, peeled, seeded and chopped	3 or 4
	freshly ground pepper	
	dry bread crumbs	

Reduce a large pinch of the mixed herbs to a powder in a mortar, sprinkle the goose with the herbs and a pinch of salt, and leave it overnight. Then wipe the goose dry.

Melt the goose fat with about 6 tablespoons [90 ml.] of water over low heat. When nothing solid is left but the cracklings, strain off the pure fat. Cook the goose gently in a deep skillet, well bathed in this fat, for one to one and a half hours, turning it every 15 minutes. Reserve the fat.

Meanwhile, put the pork rind, salt pork and pig's foot into a saucepan and cover them with cold water. Bring them to a boil, simmer them for a few minutes, then drain and rinse them in cold water. Put the beans into a heavy pan along with the carrots, onions, garlic and bouquet garni, the blanched pork products and the sausage. Pour in enough water to cover everything by about 2 inches [5 cm.], bring it slowly to a boil, and adjust the heat so that, covered, the barest simmer is maintained. Do not salt. The sausage should be removed after about 40 minutes of cooking time and reserved, and the bacon should be taken out when it is tender but still firm, after about one hour's cooking time. The pig's foot and pork rind should remain with the beans until they are done, about two hours in all. Put the rind and pig's foot aside with the sausage and pork. Discard the onion and the bouquet garni. Taste the cooking liquid and add salt if it is needed.

For the lamb stew, stew the carrots and onions and 2 tablespoons [30 ml.] of the reserved goose fat in a heavy sauté pan big enough to hold the pieces of meat placed side by side. Cook for about 15 minutes, stirring regularly, until the onions and carrots are lightly browned, then remove them, making certain to leave no fragment of onion behind.

Increase the heat, and brown the pieces of lamb in the same fat. Salt them just before turning them. When they are browned all over, sprinkle them with flour and turn them over again; return the vegetables to the pan. When the flour is lightly cooked—four or five minutes—add the wine, three of the garlic cloves and the bouquet garni. Scrape and stir the contents of the pan with a wooden spoon to loosen and dissolve frying adherents. Add the tomatoes and enough of the bean cooking liquid to cover the lamb, and cover the pan. Simmer very slowly for one and one half hours, skimming off the surface fat two or three times.

Carefully remove the pieces of meat and carrot and reserve them, discard the bouquet garni, and press the rest of the meat cooking mixture through a sieve. Return this sauce to the pan, bring it to a boil, and then set the pan halfway off the heat so that the sauce's cooking is regulated to a bubble. Cook it uncovered, skimming frequently, for 15 minutes.

To assemble the cassoulet, rub the bottom and sides of a moderately deep 2-quart [2-liter] earthenware casserole with the two remaining garlic cloves until they disintegrate completely. Untie the pork rind, cut it into small rectangles about ½ inch by 2 inches [1 cm. by 5 cm.], and distribute the rectangles over the bottom of the dish. Cut the goose into two pieces and place them on the bed of rind. Drain the beans, reserving their cooking liquid. Distribute about one third of the beans over and around the pieces of goose.

Split the pig's foot, remove and discard the largest bones, cut each half of the foot into three or four pieces, and arrange them, along with the pieces of lamb and carrot (both those from the lamb stew and from the beans), evenly over the surface. Sprinkle generously with pepper, and cover the meat and carrots with half of the remaining beans.

Slice the sausage into rounds the thickness of two fingers; distribute them throughout the casserole. Slice the bacon into 1-inch [2½-cm.] lengths and scatter them on top of the beans. Add a good pinch of pepper and cover with the remaining beans.

Generously sprinkle the top with bread crumbs and carefully, so as to moisten the bread crumbs without displacing them, ladle over them the sauce from the lamb stew. Continue until the liquid has risen just to the surface of the beans. Dust again lightly with bread crumbs, sprinkle several tablespoons of melted goose fat over the surface, and put the dish into a preheated 425° F. [220° C.] oven until it is heated through and the surface begins to bubble. Turn the oven down to 300° F. [150° C.] and maintain a gentle bubbling.

After about 20 minutes, begin to baste the surface, first with the remaining lamb sauce and, when that has been used up, with the bean cooking liquid. Continue to do this every 20 minutes or so and, when a crisp, golden gratin has formed on the surface, break it all over with a spoon so that part becomes submerged and the rest is moistened with the sauce. The cassoulet should remain in the oven at least two hours, and the gratin should be broken a minimum of three times; but if the basting liquid runs short before this time, it is better to stop the gratinéing process than to risk drying out the cassoulet.

RICHARD OLNEY
THE FRENCH MENU COOKBOOK

Cassoulet Castelnaudary-Style

Le Cassoulet de Castelnaudary

Castelnaudary, the reputed birthplace of cassoulet, lies between Toulouse and the walled city of Carcassonne in southwestern France. Rendered goose fat may be obtained by pulling the fatty materials from a goose carcass, melting them in a skillet over low heat and straining off the liquid fat that accumulates. It is also available as a canned French import.

	To serve 8	
4 cups	dried Great Northern or navy beans, soaked overnight and drained	1 liter
10 oz.	salt pork	300 g.
½ lb.	fresh pork rinds, tied in bundles	¼ kg.
1	carrot	1
3	onions, 1 stuck with 2 cloves, 2 chopped	3
2	bouquets garnis	2
5	garlic cloves	5
1½ lb.	boned pork blade steak, cut up	¾ kg.
1 lb.	boned lamb shoulder or roast, cut up	½ kg.
½ cup	rendered goose fat	125 ml.
	salt and pepper	
1¼ cups	meat stock (recipe, page 166)	300 ml.
⅓ cup	puréed tomato (recipe, page 166)	75 ml.
¼ lb.	garlic salami	125 g.
¼ lb.	Toulouse sausage	125 g.
1 cup	dry bread crumbs	¼ liter

Cook the beans in salted water with the salt pork, pork rinds, carrot, whole onion, one of the bouquets garnis and three of the garlic cloves. Skim the foam that rises to the top. The beans should be covered with the water to a depth of 1 inch [2½ cm.] and should be simmered over very low heat for about two hours, until they are just tender.

Sauté the pork and the lamb in 2 tablespoons [30 ml.] of the goose fat. Season with salt and pepper and, when the meat is browned, add the chopped onions, the other bouquet garni and the remaining garlic cloves. Moisten the sauté with the puréed tomato and a little of the stock. Cover the pan and simmer for 30 minutes, moistening the meat from time to time with more stock.

Remove the flavoring vegetables and the bouquet garni from the cooked beans. Add the sautéed meat, with its liquid, and the salami and sausage. Simmer the beans and meat together for one hour, covered. Take the pork, lamb, salami and sausage from the pan and drain them. Cut the pork and lamb into slices; skin the salami and sausage and slice them.

Layer the meats and beans in a 2-quart [2-liter] casserole, finishing with beans. Include enough of the bean cooking liquid to fill the spaces but not cover the beans. Sprinkle the top with bread crumbs and dot with the remaining goose fat. Cook uncovered in a preheated 275° F. [140° C.] oven for one and a half hours. During this time, break the gratin, or crust, at least three times (the traditional number of times is seven). After each breaking, partially submerge the gratin in the cassoulet, and baste it with cooking juices.

HISTORIA HORS SERIE: LES FRANÇAIS À TABLE

Beans, lentils and peas need boiling to destroy toxins (box, page 15).

Fruited Baked Beans with Chutney

To serve 6 to 8

2 cups	dried Great Northern or navy beans, soaked overnight	½ liter
	salt, preferably sea salt	
1½ tsp.	dry mustard	7 ml.
1	onion, finely chopped	1
½ cup	chutney, chopped	125 ml.
	freshly ground pepper	
2	apples, peeled, cored and sliced	2
3	peaches, peeled, pitted and sliced	3
½ cup	dried apricots	125 ml.
¼ to ½ cup	honey	50 to 125 ml.
¼ cup	molasses	50 ml.
½ cup	yogurt	125 ml.

Cook the beans in their soaking liquid with 1 teaspoon [5 ml.] of salt for about one hour, or until tender but not mushy. Drain the beans, reserving 1 cup [¼ liter] of the liquid.

In a small bowl, dissolve the mustard in the bean liquid and combine it with the onion and chutney; add salt and pepper to taste. Stir this mixture into the beans.

Pour half of the bean mixture into an oiled 2-quart [2-liter] casserole or baking dish. Top with about one half of the apples, peaches, and apricots; pour in the rest of the beans, and top with the rest of the fruit. Combine the honey and molasses, and pour them evenly over the top.

Cover the casserole, and bake in a preheated 325° F. [160° C.] oven for one hour; then remove the cover and bake for another 30 minutes. About five minutes before removing the casserole from the oven, pour on the yogurt. Serve steaming hot with corn bread or brown rice.

MARTHA ROSE SHULMAN
THE VEGETARIAN FEAST

Guernsey Bean Jar

To prepare honeycomb tripe, halve it, score it around the edges to allow it to lie flat, and cut it into rectangles.

Until the 1920s, Guernsey Bean Jar was the usual break-fast on the Channel island of Guernsey. Some people still indulge in this tradition.

To serve 4

2 cups	dried Great Northern or navy beans, soaked overnight and drained	½ liter
1	pig's foot	1
1	beef shank (optional)	1
10 oz.	honeycomb tripe (optional)	300 g.
4	medium-sized onions	4
8 to 10	carrots	8 to 10
1	large bouquet garni of thyme, sage and parsley	1

Place all of the ingredients in an earthenware jar or casserole and cover with water. Bring to a boil and then simmer, covered, for eight hours, adding more water when necessary.

J. STEVENS COX (EDITOR)
GUERNSEY DISHES OF BYGONE DAYS

White Beans Uccelletto

These beans are called "in the style of a little bird," and when you taste them, your spirits soar. There is a popular rhyme in Florence, which loosely translated goes, "Florentines in front of white beans lick their plates and their napkins too."

To serve 8

2 cups	dried Great Northern or navy beans, soaked overnight and drained	½ liter
3 tbsp.	dried sage leaves	45 ml.
½ cup	lightly crushed peeled tomatoes	125 ml.
2 tbsp.	puréed tomato *(recipe, page 166)*	30 ml.
¼ cup	olive oil	50 ml.
2	garlic cloves	2
	salt and pepper	
6 or 7	sweet Italian sausages	6 or 7

Place the beans in an earthenware bean pot with the sage, tomatoes and puréed tomatoes, olive oil and garlic. Add enough water just to cover. Cover the pot and set it in a preheated 350° F. [180° C.] oven. After 30 minutes, reduce the heat to 250° F. [120° C.]. From time to time, stir the beans. The cooking will take three to four hours; during the last hour, check the pot; if the beans look dry, add hot water. Add salt and pepper when the beans are almost cooked.

Place the sausages in a skillet with enough water just to cover them and boil until the water has practically evaporated. This blanching removes the fat. Add the sausages to the bean pot during the last 30 minutes of cooking.

NAOMI BARRY AND BEPPE BELLINI
FOOD ALLA FLORENTINE

White Beans in Sauce
Haricots Blancs en Sauce

To serve 10

2½ cups	dried Great Northern or navy beans, soaked overnight and drained	625 ml.
5	onions, 2 chopped and 3 thinly sliced	5
¼ tsp.	powdered saffron, steeped in 2 tbsp. [30 ml.] hot water	1 ml.
12 tbsp.	butter	180 ml.
	salt and pepper	
6 tbsp.	chopped fresh parsley	90 ml.

Place the beans in a casserole together with the chopped onions, the saffron and the butter. Add enough water to cover the beans by 1 inch [2½ cm.], cover the pan, and simmer the mixture for about one and one half to two hours, or until the beans are almost tender.

Add the sliced onions, salt and pepper to taste, and the parsley. Continue cooking just until the sliced onions are soft—about 10 minutes. If the sauce is too thick, add a little more water. Serve very hot in a deep serving dish.

AHMED LAASRI
240 RECETTES DE CUISINE MAROCAINE

Spiced White Beans
La Loubia

To serve 6

4 cups	dried Great Northern or navy beans, soaked overnight and drained	1 liter
4	garlic cloves	4
	ground cumin	
1 tsp.	Spanish paprika	5 ml.
2	whole cloves	2
	salt	
	cayenne pepper	
1 tbsp.	olive oil	15 ml.

Add enough water to the beans to cover them, and cook them over medium heat for one and one half to two hours, or until they are tender.

In a mortar, pound the garlic, a good pinch of cumin, the paprika, cloves, a little salt and a pinch of cayenne to a paste, and add the olive oil. Mix with the beans about 20 minutes before they are completely tender.

LÉON ISNARD
LA CUISINE FRANÇAISE ET AFRICAINE

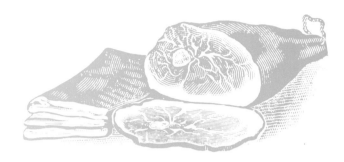

Red Bean Stew
Habichuelas Rosadas

The sweet chilies called for in this recipe are the lantern-shaped aji, available at Latin American markets.

To serve 8

2 cups	dried red beans, soaked overnight and drained	½ liter
¾ lb.	pumpkin, peeled, seeded and cut into pieces	350 g.
1 tbsp.	lard or oil	15 ml.
1 oz.	salt pork, diced	30 g.
2 oz.	lean ham, diced	60 g.
1	onion, chopped	1
1	green pepper, seeded, deribbed and chopped	1
2	sweet chilies, stemmed, seeded and chopped	2
2	garlic cloves, chopped	2
6	fresh coriander leaves, chopped	6
¼ tsp.	dried oregano leaves	1 ml.
¼ cup	tomato sauce *(recipe, page 167)*	50 ml.
1 tbsp.	salt	15 ml.

Place the beans in an 8-quart [8-liter] pot, together with the pumpkin and 2 quarts [2 liters] of water. Heat to boiling, cover, and cook over medium heat for about one hour.

Meanwhile, in a small, heavy kettle, heat the fat or oil, and rapidly brown the salt pork and ham in it. Reduce the heat to low, add the onion, green pepper, sweet chilies, garlic, coriander and oregano, and sauté the mixture for 10 minutes, stirring occasionally.

When the beans are almost tender, mash the pumpkin and add the pork mixture, tomato sauce and salt. Stir the ingredients together and simmer, uncovered, over medium heat for about one hour, or until the sauce thickens.

CARMEN ABOY VALLDEJULI
PUERTO RICAN COOKERY

Beans, lentils and peas need boiling to destroy toxins (box, page 15).

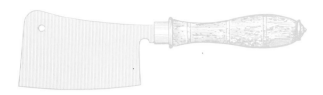

Savory Beans

Frijoles a la Charra

The serrano chili used in this recipe is a small smooth green chili, usually rounded at the end. It has a strong, fresh flavor, and its seeds and veins are very hot. It is available at Latin American markets. The volatile oils in hot chilies may make your skin sting and your eyes burn; after handling chilies, avoid touching your face and wash your hands promptly.

The Mexican title of this dish means, literally, beans cooked in the way a lady *charro* would prepare them. (The *charros* are the elegant horsemen of Mexico.) The green chilies and coriander leaves give the beans a unique and interesting flavor, making them a perfect complement to simple charcoal-grilled meats.

To serve 6

1 cup	dried pink or pinto beans	¼ liter
¼ lb.	fresh pork rind, cut into small squares	125 g.
¼	onion, sliced	¼
2	small garlic cloves, finely chopped	2
1¼ tsp.	salt	6 ml.
¼ lb.	bacon, cut in small pieces	125 g.
2 tbsp.	lard or pork drippings	30 ml.
2	medium-sized tomatoes, peeled, seeded and chopped, juice strained and reserved	2
3	serrano chilies, stemmed and finely chopped	3
2	large sprigs coriander	2

Put the beans, pork-rind squares, onion and garlic into a bean pot or large saucepan. Add 6 cups [1½ liters] of water and bring it to a boil. Reduce the heat, cover the pot, and let the beans cook gently until they are tender, about one and one half hours. Add the salt and cook the beans, uncovered, for another 15 minutes.

In a separate pan, cook the bacon gently in the lard or drippings until it is slightly browned. Add the tomatoes with their juice, and stir in the chilies and coriander. Cook the tomato mixture over a fairly high heat for about 10 minutes, until it is well seasoned and thick.

Add the tomato mixture to the beans and let everything cook, uncovered, over low heat for about 15 minutes.

Serve in small individual bowls with grilled meat, or add some more liquid and serve as a soup.

DIANA KENNEDY
THE CUISINES OF MEXICO

Kidney Beans in Walnut Sauce

Lobio

This is a thoroughly Georgian combination that is usually offered as one of several appetizers. Instead of the herbs suggested below, a mixture of ½ tablespoon [7 ml.] each of finely chopped fresh coriander, parsley, mint and basil leaves may be used. The dish may be accompanied by a platter of trimmed scallions and/or a bowl of whole fresh coriander, parsley, mint and basil leaves.

To serve 4

1 cup	dried red kidney beans, soaked overnight and drained	¼ liter
	salt	
1 oz.	walnuts (about ¼ cup [50 ml.])	30 g.
1	small garlic clove	1
	cayenne pepper	
1 tbsp.	wine vinegar, mixed with 3 tbsp. [45 ml.] water	15 ml.
2 tbsp.	finely chopped onion	30 ml.
2 tbsp.	finely chopped fresh coriander leaves	30 ml.
2 tbsp.	finely chopped fresh parsley	30 ml.

Put the beans in a saucepan, cover them with fresh water, and cook, uncovered, for about one and one half to two hours, or until the beans are tender but still intact. Add boiling water as needed to keep the beans immersed during cooking. When the beans are done, drain them, place them in a bowl and season them with salt.

In a mortar, pound the walnuts to a paste with the garlic and a pinch of cayenne. Stir in the vinegar and water until well blended. Add the walnut mixture, along with the onion, coriander and parsley, to the beans. Mix well, being careful not to bruise the beans. Cover and chill before serving.

SONIA UVEZIAN
THE BEST FOODS OF RUSSIA

Red-hot Bean Casserole

The volatile oils in hot chilies may make your skin sting and your eyes burn; avoid touching your face and wash your hands promptly.

Savory brown rice is an almost essential accompaniment to this dish. If you cannot buy hot chilies, use ¼ teaspoon [1 ml.] of cayenne pepper.

	To serve 4	
1 cup	dried red kidney beans, soaked overnight and drained	¼ liter
4	medium-sized tomatoes, peeled and sliced	4
2	medium-sized onions, thinly sliced	2
1	large garlic clove, finely chopped	1
2	large red peppers, seeded, deribbed and cut into julienne	2
2	fresh hot red chilies, stemmed, seeded and finely chopped	2
2 tsp.	paprika	10 ml.

Place the beans in a large pot, cover them with water, bring to a boil, cover the pot and cook the beans for one hour. Drain the beans.

Put half of the tomatoes in the bottom of a casserole. Add half of the onions and garlic, half of the red peppers, half of the chopped chilies and half of the paprika.

Put in all of the beans and follow with the remaining ingredients, in layers in reverse order, topping the layers with the tomatoes. Cover the casserole and put it into a preheated 325° F. [160° C.] oven for one and one half hours. Stir the layers to mix the ingredients before serving.

GAIL DUFF
GAIL DUFF'S VEGETARIAN COOKBOOK

Bean Croquettes

The cooking of kidney beans is explained on pages 16-17.

	To serve 6	
2 cups	cooked red or white kidney beans	½ liter
4 tbsp.	butter	60 ml.
¼ tsp.	ground bay leaf	1 ml.
1 tsp.	vinegar	5 ml.
	salt and pepper	
1	egg	1
	fine dry bread crumbs	
	oil for deep frying	

Purée the cooked beans through a fine sieve. Combine the bean purée with the butter, bay leaf and vinegar. Season with salt and pepper. Blend the purée thoroughly, then whip it until it is light. Form the purée into 12 small cylinders.

Beat the egg with a little water and dip the croquettes in it. Dredge the croquettes in the bread crumbs to coat them evenly all over. Repeat the process. Arrange the croquettes on a platter without letting them touch each other. Chill them for about one hour. Just before serving, deep fry the croquettes a few at a time in oil heated to 375° F. [190° C.]. Drain the croquettes on paper towels and serve them as soon as possible.

MORTON G. CLARK
FRENCH-AMERICAN COOKING

Kidney Bean Purée

The beans can be completely cooked and puréed in advance and kept refrigerated. Reheat at serving time with a little red wine added to prevent scorching. For a different treat, the purée can be formed into patties, coated with flour and fried in butter.

	To serve 6	
1½ cups	dried red kidney beans, soaked overnight and drained	375 ml.
	salt	
1	bay leaf	1
1	carrot, cut in half lengthwise	1
1	small onion, stuck with 1 whole clove	1
1½ cups	red wine	375 ml.
¼ tsp.	freshly grated nutmeg	1 ml.
	pepper	
3 tbsp.	butter (optional)	45 ml.

Place the beans in a large pot, and pour in enough water to cover them by 1 inch [2½ cm.]. Add 1 teaspoon [5 ml.] of salt, the bay leaf, the carrot and the onion. Cover and cook until the beans are tender—about one and one half hours.

Drain the beans. Discard the bay leaf and the clove from the onion. Pour half of the wine into a blender, and purée half of the beans with the onion and the carrot. Repeat with the remaining wine and beans.

Return the purée to the bean pot, add the nutmeg, season with salt and pepper, and add the butter if you are using it. Reheat the purée and simmer it for five minutes. Spoon the purée onto a platter around carved meat or pass it in a separate vegetable bowl.

CAROL CUTLER
THE SIX-MINUTE SOUFFLÉ AND OTHER CULINARY DELIGHTS

Beans, lentils and peas need boiling to destroy toxins (box, page 15).

Bean Polenta

Polenta d'Haricots

To serve 6 to 8

2 cups	dried red or navy beans, soaked overnight and drained	½ liter
1 tbsp.	molasses	15 ml.
½ tbsp.	prepared mustard	7 ml.
1 tbsp.	butter	15 ml.
1 tbsp.	vinegar	15 ml.
3 tbsp.	strained fresh lemon juice	45 ml.
	salt and pepper	

Cover the beans with cold water, bring the water to a boil, and cook the beans until tender—one and one half to two hours. Pour the cooked beans into a sieve and press them through. Put the resulting pulp into a pan and stir in the molasses, mustard, butter, vinegar, lemon juice, and salt and pepper to taste. Gently reheat before serving.

THE ORIGINAL PICAYUNE CREOLE COOK BOOK

Lamb Neck and Bean Casserole

This dish takes a long time to make, but it can be assembled in stages or made completely in advance and reheated.

To serve 6

2 cups	dried red or white kidney beans, soaked overnight and drained	½ liter
1½ lb.	lamb neck, cut into small pieces	¾ kg.
¼ cup	oil	50 ml.
2	medium-sized onions, sliced	2
2	garlic cloves, finely chopped	2
3	small tomatoes, peeled and chopped	3
½ tsp.	dried thyme leaves	2 ml.
1	bay leaf	1
½ tsp.	crushed dried hot chili (optional)	2 ml.
1 cup	dry red wine	¼ liter

Cover the beans with water and cook for one and one half to two hours. Drain the beans, reserving the cooking liquid.

Dry the lamb pieces with paper towels. Heat the oil in a heavy skillet. Toss in the lamb and brown the pieces over medium heat. Transfer the lamb to a casserole and season it lightly with salt and pepper.

Sauté the onions and garlic over low heat in the skillet used to brown the lamb, stirring to scrape up the browned bits and juices of the lamb. Add the tomatoes, herbs and wine, and bring the mixture to a boil, stirring constantly; simmer for a few minutes.

Gently mix the beans into the lamb, taking care not to mash the beans. Pour the hot tomato and wine sauce over the contents of the casserole; if the sauce does not cover the lamb and beans completely, add enough bean liquid to take care of that. Put a lid on the casserole, or make one out of a double thickness of aluminum foil, sealing it tightly at the edges.

Put the casserole in a preheated 325° F. [170° C.] oven for about one hour. In this time, the beans should absorb most of the liquid; if the mixture seems too juicy, uncover the casserole, turn up the heat to 375° F. [190° C.], and bake until the excess moisture evaporates.

MIRIAM UNGERER
GOOD CHEAP FOOD

Lima Bean Casserole

The brown—unhulled—sesame seeds called for in this recipe are available at health-food stores.

To serve 4 to 6

⅔ cup	dried lima beans, soaked overnight and drained	150 ml.
2 tbsp.	oil	30 ml.
1	large green pepper, seeded, deribbed and diced	1
2	celery ribs, diced	2
1	large onion, chopped	1
2	garlic cloves, finely chopped	2
1½ tsp.	salt	7 ml.
1½ cups	chicken stock *(recipe, page 166)* (optional)	375 ml.
1½ tbsp.	brown sesame seeds	22 ml.
½ cup	raisins	125 ml.
6 oz.	Cheddar cheese, shredded (about 1½ cups [375 ml.])	175 g.

Cook the beans in enough water to cover them for one and one half to two hours, or until they are tender. Drain and reserve the bean cooking liquid. Heat the oil and sauté the green pepper, celery, onion and garlic in it until they are tender—about 10 minutes. Add to the sautéed vegetables the cooked lima beans, salt, sesame seeds, raisins, and either the bean cooking liquid or chicken stock. Mix the ingredients well, then add half of the cheese. Turn the mixture into an oiled casserole and bake in a preheated 375° F. [190° C.] oven for 25 minutes. Sprinkle on the remaining cheese, increase the heat to 400° F. [200° C.], and bake for 15 minutes to form a golden crust.

NANCY ALBRIGHT
RODALE'S NATURALLY GREAT FOODS COOKBOOK

Mung Beans with Pork and Shrimp

Gulay Na Mongo

To serve 4

1 cup	dried mung beans, soaked for 1 hour and drained	¼ liter
2	garlic cloves, chopped	2
2 tbsp.	lard	30 ml.
1	small onion, chopped	1
5 oz.	pork shoulder, cut into small pieces	150 g.
¼ cup	tomato juice	50 ml.
5 oz.	large fresh shrimp, shelled, deveined, halved lengthwise and then crosswise	150 g.
	salt	

Boil the beans in 4 cups [1 liter] of water for 40 to 60 minutes, or until tender. Drain the beans and keep them hot. Sauté the garlic in the lard and add the onion, stirring until the onion is transparent. Add the pork and cover the pan. Stir occasionally until the pork is well done—about 30 minutes. Pour in the tomato juice. Add the shrimp and cook, covered, for 10 minutes. Finally, add the hot mung beans. Stir thoroughly and add salt to taste.

ALICE MILLER MITCHELL (EDITOR)
ORIENTAL COOKBOOK

Mung Beans with Capers and Lemon

To serve 4

1 cup	dried mung beans, soaked for 1 hour	¼ liter
¼ cup	olive oil	50 ml.
2	medium-sized onions, finely chopped	2
1	garlic clove, chopped	1
2 tbsp.	capers, chopped	30 ml.
1	large lemon, the peel grated and the juice strained	1
¼ cup	chopped fresh parsley	50 ml.

In a covered pot, cook the mung beans in the soaking water for 40 minutes, then drain them if necessary. Heat the oil in a saucepan over low heat. Mix in the onions and garlic and cook them until they are soft. Stir in the beans, capers and the peel and juice of the lemon. Cover and simmer for two minutes. Mix in the parsley just before serving.

GAIL DUFF
GAIL DUFF'S VEGETARIAN COOKBOOK

Gratinéed Beans and Tomatoes

Bohnenkerne

To serve 4

2 cups	dried navy beans, soaked overnight and drained	½ liter
1	bay leaf	1
	salt	
4 tbsp.	butter	60 ml.
2 oz.	bacon, diced	60 g.
1	onion, sliced	1
3	medium-sized tomatoes, sliced	3
½ cup	shredded Gruyère cheese	125 ml.

Put the beans and bay leaf in a pan; cover them with water. Cook the beans, covered, for two hours, or until tender. Salt the beans to taste toward the end of the cooking. Melt half of the butter, and in it lightly brown the bacon and onion. Fill a greased baking dish with layers of the beans, tomato slices, and bacon and onion. Dot with the remaining butter, sprinkle on the cheese, and bake in a preheated 350° F. [180° C.] oven for one hour, or until the cheese is browned.

ELIZABETH SCHULER
MEIN KOCHBUCH

Boston Baked Beans

To serve 6 to 8

2½ cups	dried navy or kidney beans, soaked overnight and drained	625 ml.
6 tbsp.	brown sugar	90 ml.
½ cup	molasses	125 ml.
2 tsp.	dry mustard	10 ml.
2 tsp.	salt	10 ml.
1	medium-sized onion	1
6 oz.	fat salt pork, cut into 3 pieces	175 g.

Put the beans in a saucepan with enough cold water to cover them, bring to a boil and drain them, saving the water. Transfer the beans to a 6-cup [1½-liter] bean pot or deep casserole. Stir in the brown sugar, molasses, mustard and salt. Poke the onion down into the center. Stick two pieces of the pork down into the beans and place the third piece on top. Add enough of the reserved water to cover the beans. Cover the bean pot and bake all day, usually seven to eight hours, at 275° F. [140° C.]. Lift the lid occasionally to be sure the beans are still moist, adding more of the reserved water if necessary. Uncover the pot for the last hour of cooking. Remove the onion before serving the beans.

CHARLOTTE TURGEON AND FREDERIC A. BIRMINGHAM
THE SATURDAY EVENING POST ALL AMERICAN COOKBOOK

Beans, lentils and peas need boiling to destroy toxins (box, page 15).

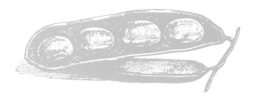

Dry Bean Torte

Bohnen Torte

To make vanilla sugar, place sugar and a vanilla bean in a closed container for about a week. Remove the bean and save it for a later use. If you like, this torte can be made with regular sugar and ½ teaspoon [2 ml.] of vanilla extract added to the batter.

To serve 10 to 12

¾ cup	small dried navy beans, soaked overnight and drained	175 ml.
3	eggs, the yolks separated from the whites	3
¾ cup	vanilla sugar	175 ml.
2 tbsp.	rum	30 ml.
¼ lb.	finely ground walnuts or hazelnuts (about 1 ¼ cups [300 ml.])	125 g.
3 to 4 tbsp.	apricot jam	45 to 60 ml.
	confectioners' sugar	

Cook the beans, well covered with water, over medium heat for about one hour, or until tender. Drain off the water and steam the beans over low heat to remove all excess dampness. Rice the beans through a food mill.

Beat the egg yolks with half of the sugar until light and fluffy. Gradually add the riced beans and the rum. Mix the batter until it is creamy and very well blended. Add the nuts.

Beat the egg whites until soft peaks form. Spoon the remaining sugar gradually into the egg whites, beating after each addition, until the whites become very stiff. Fold the whites gently into the batter and pour the batter into a well-buttered and floured 9-inch [23-cm.] spring-form pan.

Bake in a preheated 350° F. [180° C.] oven for about one hour, until the torte is firm and shrinks from the sides of the pan. Next day slice the torte into two layers; spread apricot jam on one layer and top with the second layer. Dust with confectioners' sugar and serve.

LILLY JOSS REICH
THE VIENNESE PASTRY COOKBOOK

T-Dart Pinto Beans

This recipe comes from the T-Dart Ranch in Phoenix, Arizona. The volatile oils in hot chilies may make your skin sting and your eyes burn; after handling chilies, avoid touching your face and wash your hands promptly.

To serve 8

5 cups	dried pinto beans, soaked overnight and drained	1 ¼ liters
2 lb.	ground beef	1 kg.
2	medium-sized onions, chopped	2
1	garlic clove, crushed	1
	salt	
	dried oregano leaves	
1	dried hot red chili, finely crushed	1
4 tbsp.	rendered bacon fat	60 ml.

Cover the beans with fresh cold water and bring them to a boil. Reduce the heat to low, cover the pot, and allow the beans to cook slowly.

Mix the meat with the onions, garlic, 2 teaspoons [10 ml.] salt, a pinch of oregano and the chili. Heat the bacon fat in a skillet. Add the meat mixture and fry, stirring frequently to break up lumps, until the meat is golden brown—about 10 minutes. Add the meat to the beans and simmer, partially covered, for four hours. Add more salt to taste.

CLEMENTINE PADDLEFORD
THE BEST IN AMERICAN COOKING

Cowpoke Beans

The volatile oils in hot chilies may make your skin sting and your eyes burn; after handling chilies, avoid touching your face and wash your hands promptly.

To serve 6 to 8

2 cups	dried pinto or red kidney beans, soaked overnight	½ liter
1	ham bone	1
1	fresh hot red chili (optional)	1
1 tsp.	salt	5 ml.
¼ lb.	suet, chopped	125 g.
1	large onion, chopped	1
1	garlic clove, chopped	1
4	ripe tomatoes, chopped	4
6 tbsp.	chopped fresh parsley	90 ml.
½ tsp.	ground cumin	2 ml.
½ tsp.	ground marjoram	2 ml.
1 ½ tbsp.	chili powder	22 ml.

Put the beans and the water in which they soaked into a saucepan. Add the ham bone, the hot chili, if using, and the salt. Bring to a boil, reduce the heat, cover, and simmer gently for one and one half to two hours, or until the beans are tender. Drain the beans and reserve the cooking liquid.

Heat the suet in a large skillet, stir in the onion and the garlic, and cook over medium heat for five minutes, until the onion takes on a little color. Add the tomatoes, parsley, 1 cup [¼ liter] of the bean liquid and all of the remaining ingredients. Cook this sauce over a low heat, stirring frequently, for 45 minutes. Add the beans, cover, and continue simmering gently for 20 minutes, until all of the ingredients are heated through and the flavors are well blended.

THE EDITORS OF AMERICAN HERITAGE
THE AMERICAN HERITAGE COOKBOOK

Chili con Carne

To serve 8

2 cups	dried pinto, red kidney or similar beans, soaked overnight and drained	½ liter
2 lb.	lean stewing beef, cut into cubes	1 kg.
2	bay leaves	2
2	large onions, sliced	2
1	garlic clove, finely chopped	1
2 tbsp.	oil	30 ml.
5	tomatoes, peeled, seeded and chopped, or 3 cups [¾ liter] tomato sauce *(recipe, page 167)*	5
2 tsp.	salt	10 ml.
1 tbsp.	cornstarch or 2 tbsp. [30 ml.] cornmeal	15 ml.
¼ tsp.	dried oregano leaves	1 ml.
¼ tsp.	dried sage leaves	1 ml.
¼ tsp.	ground cumin	1 ml.
1½ tsp.	pepper	7 ml.
1 tbsp.	chili powder	15 ml.

Put the beans and the meat in enough water to cover them, and bring to a boil. Add the bay leaves, onion and garlic, and continue cooking over moderate heat until the beans are tender—about two hours.

Heat a skillet, then add the oil. Stir in the tomato, salt, cornstarch or cornmeal, herbs and seasonings. Mix them thoroughly and simmer for five minutes. Add the mixture to the beans and simmer for one more hour.

GEORGE C. BOOTH
THE FOOD AND DRINK OF MEXICO

Casserole of Duck with Pinto Beans

To serve 6

2 cups	dried pinto beans, soaked overnight and drained	½ liter
2	garlic cloves	2
1	bay leaf	1
3	onions, 1 stuck with 2 cloves, 2 finely chopped	3
two 5 lb.	ducks, cut into serving pieces, skin removed and reserved	two 2½ kg.
about 1 cup	flour	about ¼ liter
	salt and freshly ground pepper	
2 tbsp.	butter	30 ml.
2 oz.	salt pork, diced	60 g.
	dried basil	
½ tsp.	dry mustard	2 ml.
	oil for deep frying	

Place the beans in a pot and add the garlic, bay leaf and the whole onion. Cover with boiling salted water and cook over medium heat until the beans are tender—one and one half to two hours. Drain them, reserving the liquid.

Cut the skin into thin strips and set them aside. Dredge the duck pieces in the flour, seasoned with salt and pepper.

Melt the butter in a skillet over medium heat, and fry the salt pork until most of its fat is rendered. Add the chopped onion and sauté until the onion is limp. Transfer the pieces of salt pork and onion to a casserole.

Sear the duck in the salt-pork fat and transfer it to the casserole. Add the beans, a pinch of basil, some additional pepper and the dry mustard. Mix the duck with the beans. Pour in enough bean liquid to cover the ingredients, and simmer the mixture, covered, in a preheated 350° F. [180° C.] oven until the beans and duck are thoroughly tender—about two hours. Add liquid if it cooks away too quickly.

Meanwhile, deep fry the strips of skin in oil heated to 350° F. [180° C.] for two to three minutes to crisp. Drain the strips and use them as a garnish for the duck.

JAMES BEARD
JAMES BEARD'S FOWL & GAME BIRD COOKERY

Beans, lentils and peas need boiling to destroy toxins (box, page 15).

Soybeans and Vegetables in Sour Cream-Black Mushroom Sauce

Soy Gevult

For the cooked vegetables, steam up some of your favorites: carrot, cabbage, zucchini—whatever. Recommended amount, about 6 cups [1½ liters] worth—one small head cabbage, one medium sliced carrot, one smallish zucchini.

To serve 6 to 8

2 cups	dried soybeans, soaked for 3 hours and drained	½ liter
1 cup	sour cream	¼ liter
½ cup	Chinese dried black mushrooms, soaked in 2½ cups [625 ml.] warm water for 20 minutes, drained and stems cut off, the soaking liquid reserved	125 ml.
3 tbsp.	butter	45 ml.
1½ cups	chopped onion	375 ml.
2	garlic cloves, crushed	2
1 tsp.	salt	5 ml.
3 tbsp.	flour	45 ml.
¼ cup	sherry	50 ml.
¼ cup	*tamari* soy sauce	50 ml.
	assorted cooked vegetables	
	black pepper	

Cover the soybeans with water, add salt and cook for one and three quarters hours, or until the beans are tender. Drain the beans and, while they are still hot, toss them with the sour cream. Drain the mushrooms thoroughly, squeezing out excess water. Reserve the liquid; it holds flavor secrets. Slice the mushrooms thin.

Melt the butter in a saucepan. Add the onion, garlic, sliced mushrooms and salt, and cook over medium heat for five minutes; stir in the flour. Keep stirring and cook for five more minutes.

Add the mushroom water and 2 tablespoons [30 ml.] each of the sherry and *tamari* sauce. Reduce the heat to very low and cook, stirring frequently, for eight to 10 minutes, or until the sauce is thickened and smooth. Combine the beans, sauce and the vegetables in a large casserole. Add the remaining sherry and *tamari* sauce and lots of black pepper. Bake the casserole, covered, in a preheated 350° F. [180° C.] oven for about 45 minutes.

MOLLIE KATZEN
THE MOOSEWOOD COOKBOOK

Bean Curd

Stewed Bean Curd

For a chewier, more meaty texture, freeze the bean curd before adding it to the simmering liquid.

To serve 6

2	cakes bean curd *(recipe, page 165)*, sliced into 16 equal pieces	2
½ cup	oil	125 ml.
2	onions, sliced	2
2	garlic cloves, finely chopped	2
2 or 3	carrots, coarsely chopped	2 or 3
½ lb.	cabbage, coarsely chopped	¼ kg.
½ cup	soy sauce	125 ml.

Heat the oil in a 4-quart [4-liter] casserole or a cast-iron Dutch oven; sauté the onions and then the garlic until they are soft but not browned. Add the carrots and cabbage, and sauté for one more minute, then add the bean curd and cook for one or two additional minutes. Add the soy sauce and 1 cup [¼ liter] of water, and gently boil the mixture for about 15 to 20 minutes, until the juices thicken slightly.

TOM RIKER & RICHARD ROBERTS
THE DIRECTORY OF NATURAL & HEALTH FOODS

Fried Bean Curd

To serve 4

2	cakes bean curd *(recipe, page 165)*, cubed	2
2 tbsp.	oil	30 ml.
½ tsp.	peeled and grated fresh ginger	2 ml.
1	scallion, finely chopped	1
1 tsp.	salt	5 ml.
1 tbsp.	rice wine or dry sherry	15 ml.
2 tsp.	cornstarch mixed with 2 tbsp. [30 ml.] cold chicken stock *(recipe, page 166)*	10 ml.
1 tsp.	Chinese sesame-seed oil	5 ml.

Heat up a wok or skillet over high heat before putting in the oil. Add the ginger and scallion, followed almost immediately by the bean curd. Stir fry for two to three minutes, breaking up the bean curd into even smaller bits.

Add the salt and wine together with the cornstarch mixed with the chicken stock. Cook, stirring often, until the sauce thickens slightly. Add the sesame-seed oil and serve.

DEH-TA HSIUNG
CHINESE REGIONAL COOKING

Mrs. Chen's Bean Curd

Ma-po Dou-fu

Szechwan hot bean paste, Szechwan peppercorns, rice wine, fermented salted black beans and tree-ear mushrooms are all sold at Asian food markets.

To serve 4

2	cakes firm bean curd (recipe, page 165), cubed	2
1 to 2 tsp.	salt	5 to 10 ml.
1 tbsp.	rice wine or dry sherry	15 ml.
1 tbsp.	soy sauce	15 ml.
about ½ tsp.	ground Szechwan peppercorns	about 2 ml.
2 to 3 tbsp.	cornstarch, dissolved in 4 to 6 tbsp. [60 to 90 ml.] water	30 to 45 ml.
4 to 5 tbsp.	oil	60 to 75 ml.
⅓ to ½ lb.	fatty pork, diced	150 to 250 g.
1 to 2 tbsp.	Szechwan hot bean paste	15 to 30 ml.
3 to 5 tbsp.	peeled and finely grated fresh ginger	45 to 75 ml.
2 tbsp.	finely chopped garlic (optional)	30 ml.
1 tsp.	fermented salted black beans (optional)	5 ml.
1 cup	chicken stock (recipe, page 166) or water	¼ liter
2 or 3	fresh mushrooms, or dried mushrooms soaked in warm water for 20 minutes (optional)	2 or 3
6 to 8	pieces tree ears, soaked in warm water until softened, about 30 minutes, stems removed (optional)	6 to 8
1 tbsp.	Chinese sesame-seed oil (optional)	15 ml.
¼ cup	coarsely chopped scallions	50 ml.

Mix the salt, rice wine or dry sherry, soy sauce and Szechwan pepper into the dissolved cornstarch. Heat the oil in a wok or large frying pan until very hot. Add the pork pieces and cook briefly. Stir in the hot bean paste, then the ginger and, if using, the garlic and the fermented black beans. Stir the mixture until the meat and ginger have absorbed the red color from the hot bean paste. Add the stock or water. Then carefully add the bean-curd cubes and stir gently.

Keeping the wok over medium heat, let the liquid cook down somewhat. Stir occasionally, being careful not to break the bean-curd cubes. Just before the liquid has cooked away, stir in the seasoned cornstarch mixture and, if you are including them, the mushrooms, tree ears and sesame-seed oil. Add the scallions at this time or at any time until just before serving, depending on whether you want the onion flavor to be cooked into the sauce or whether you want to leave the onion flavor more isolated and distinct. Stir, check seasoning for salt, and check the consistency of the sauce—it should now be very thick, almost custard-like; if it is too thin, add a tablespoon or so more cornstarch, first mixing it with a few tablespoons of cold water to make a thin paste. Sprinkle additional ground Szechwan peppercorns over the bean curd. Serve hot—on top of rice, if you like.

ROBERT A. DELFS
THE GOOD FOOD OF SZECHWAN

Sizzling Bean Curd with Ginger Ankake Sauce

To serve 4

1½	cakes bean curd (recipe, page 165)	1½
⅓ cup	cornstarch	75 ml.
2	egg whites	2
	oil for deep frying	
Ginger ankake sauce		
⅔ cup	vegetable stock (recipe, page 166) or water	150 ml.
3 tbsp.	soy sauce	45 ml.
2 tbsp.	brown sugar or 1 tbsp. [15 ml.] honey	30 ml.
2 tsp.	cornstarch, dissolved in 2 tbsp. [30 ml.] water	10 ml.
2 tsp.	peeled and grated fresh ginger	10 ml.

To prepare the sauce, combine the stock or water, soy sauce and sugar in a small saucepan and bring to a boil. Stir in the dissolved cornstarch and the ginger, and cook for about one minute more until thick. Keep the sauce warm while you fry the bean curd.

Combine the cornstarch and egg whites in a small bowl; mix until smooth. Cut the bean curd into 12 equal cubes or thin rectangles, dip them into the batter, and deep fry them in oil heated to 375° F. [190° C.] until golden brown. Drain them briefly on a wire rack, then transfer onto absorbent paper. Serve them immediately, topped with the hot sauce.

WILLIAM SHURTLEFF & AKIKO AOYAGI
THE BOOK OF TOFU

Beans, lentils and peas need boiling to destroy toxins (box, page 15).

Bean Curd and Vegetable Tempura

The technique of deep frying bean curd is demonstrated on pages 50-51.

To serve 4

1	cake firm bean curd *(recipe, page 165)*	1
6	fresh mushrooms	6
24	green beans	24
½	sweet potato or yam, cut into ½-inch [1-cm.] rounds	½
1	onion, cut into ½-inch [1-cm.] rounds	1
2	green peppers, seeded, deribbed and quartered lengthwise	2
¼	celeriac or 1 crisp green apple, cut into ½-inch [1-cm.] slices	¼
	oil for deep frying	
	Tempura batter	
1	egg yolk or whole egg	1
1¼ to 1½ cups	unbleached flour	300 to 375 ml.
½ tsp.	salt (optional)	2 ml.
	Tempura dipping sauce	
1 cup	vegetable stock *(recipe, page 166)*	¼ liter
3 to 4 tbsp.	rice wine or pale sherry	45 to 60 ml.
¼ cup	soy sauce	50 ml.
4 to 6 tbsp.	peeled and grated fresh ginger	60 to 90 ml.
4 to 6 tbsp.	grated white radish	60 to 90 ml.

Cut the cake of bean curd crosswise into six pieces, and press them down with a plate for several minutes. To rid the vegetables of any excess moisture, pat their cut surfaces lightly with a dry cloth.

To prepare the dipping sauce, combine all of the sauce ingredients in a saucepan, bring to a boil and remove from the heat. Let the sauce cool before serving.

Pour the oil into a wok, skillet or deep fryer and set the pan over low heat. Meanwhile, quickly prepare the tempura batter: In a mixing bowl, combine 1 cup [¼ liter] of ice water and the egg yolk or egg; beat well with a wire whisk or chopsticks. Sprinkle the flour and the salt, if used, evenly over the mixture. With a few quick strokes of the whisk or a wooden spoon, lightly stir in the flour until all of the flour is moistened and large lumps disappear—the presence of small lumps is all right. Do not stir the batter again after the initial mixing. Use it as soon as possible and do not place it too close to the heat.

When the oil temperature reaches 350° F. [180° C.], dip the bean-curd pieces into the batter and slide them into the oil. Deep fry them on both sides until golden brown, drain briefly on a wire rack, then transfer them to paper towels. Dip the vegetable slices into the batter, and deep fry six to eight at a time until golden brown.

Arrange the bean curd and vegetable tempura in individual portions on top of fresh sheets of white paper placed on serving plates or small bamboo trays. Serve accompanied by the dipping sauce.

WILLIAM SHURTLEFF & AKIKO AOYAGI
THE BOOK OF TOFU

Red-cooked Pork Chops with Bean Curd

Hung Sao Tsu Pai Dow Fu

Cloud ears are large, thick, dried mushrooms similar to the somewhat more common, though smaller, tree ears. Both are obtainable at Asian food markets.

To serve 6

3	cakes bean curd *(recipe, page 165)*, cubed	3
3 tbsp.	peanut oil	45 ml.
2	scallions, sliced	2
2	slices peeled fresh ginger, slivered	2
2	pork chops, sliced into thin strips (about ⅔ lb. [300 g.])	2
6	Chinese dried black mushrooms, soaked in 1 cup [¼ liter] warm water for 15 minutes, drained and sliced, water reserved	6
12	cloud ears, soaked in cold water for 15 minutes, rinsed and drained (optional)	12
2 tbsp.	soy sauce	30 ml.
½ tsp.	sugar	2 ml.

Put the oil into a very hot skillet and heat it until it begins to smoke. Fry the ginger and scallions for 30 seconds in the hot oil. Add the pork and black mushrooms and cook until the pork turns white—about one minute.

If you are using the cloud ears, cut each one into 16 pieces and add the pieces to the pork mixture. Add the bean curd, soy sauce, sugar and ½ cup [125 ml.] of the mushroom soaking water. Cook for five minutes and serve hot.

WONONA W. AND IRVING B. CHANG
HELENE W. AND AUSTIN H. KUTSCHER
AN ENCYCLOPEDIA OF CHINESE FOOD AND COOKING

Home-Style Bean Curd

Yellow-bean sauce is a thick sauce made from yellow beans, flour and water. It is sold at Chinese food markets. The volatile oils in hot chilies may make your skin sting and your eyes burn; after handling chilies, avoid touching your face and wash your hands promptly.

	To serve 6	
3	cakes bean curd *(recipe, page 165)*	3
	oil for deep frying	
¼ lb.	pork, cut in thin slivers	125 g.
5 or 6	dried hot red chilies, coarsely chopped	5 or 6
1 tbsp.	rice wine or sherry	15 ml.
1 tbsp.	soy sauce	15 ml.
1	leek, cut diagonally into chunks	1
2 tbsp.	yellow-bean sauce, crushed	30 ml.
½ tsp.	salt	2 ml.
½ tsp.	Chinese sesame-seed oil	2 ml.

Slice each cake of bean curd into three or four thin rectangles, then cut each rectangle diagonally into two triangles. Heat the oil in a wok to 375° F. [190° C.], and deep fry the bean-curd pieces for about two minutes; drain. Pour most of the oil out of the wok, leaving about 1 tablespoon [15 ml.]. Put in the pork and chilies and stir. Add the wine, soy sauce, bean curd, leek, bean sauce and salt, and cook the mixture for about three minutes. Add the sesame-seed oil and serve.

DEH-TA HSIUNG
CHINESE REGIONAL COOKING

Lentils with Chorizo

Lentejas con Chorizo

	To serve 4	
2½ cups	dried lentils	625 ml.
3	onions	3
2	garlic cloves	2
1	bay leaf	1
2	potatoes, peeled and quartered	2
¼ lb.	*chorizo*, cut into 2-inch [5-cm.] pieces	125 g.
	salt	

Put the lentils in a pot and add enough water to cover them by 1 inch [2½ cm.]. Add the onions, garlic and bay leaf. Cover and bring to a boil over medium heat. Add the potatoes, *chorizo* and salt. Reduce the heat and simmer, covered, until the vegetables and lentils are tender—about 30 minutes.

DONN E. POHREN
ADVENTURES IN TASTE

Lentils with Pumpkin and Dried Beef

The original recipe for this dish called for khlii, a Moroccan sun-dried beef spiced with coriander, cumin and garlic. It is difficult to obtain in the United States, but a good substitute is bastourma, a heavily spiced, dried beef obtainable at Middle Eastern or Greek food markets.

The volatile oils in hot chilies may make your skin sting and your eyes burn; after handling chilies, avoid touching your face and wash your hands promptly.

	To serve 8	
2½ cups	dried lentils	625 ml.
2 tbsp.	fat or lard	30 ml.
2	onions, sliced	2
1 or 2	dried hot red chilies or 2 or 3 small dried hot green chilies, crushed	1 or 2
1½ tbsp.	paprika	22 ml.
½ lb.	*bastourma*, in 1 piece	¼ kg.
	salt	
2 lb.	pumpkin, peeled, seeded and cubed	1 kg.
3	small tomatoes, peeled, seeded and chopped	3

Melt the fat or lard in a large, heavy-bottomed pot and cook the onions together with the chilies, paprika and *bastourma* until the onions are wilted. Add 2 quarts [2 liters] of water and season the mixture with salt. Bring the mixture to a boil; reduce the heat to low.

In another large pot, parboil the lentils for one to two minutes in boiling water. Drain them, and pour them immediately into the pot with the meat. When the lentils are almost cooked—after 15 minutes—add the pumpkin and tomatoes. Take the stew off the heat as soon as the pumpkin is cooked—about 20 minutes. Remove the meat, slice it thin and put it back into the pot. Add a little water to the stew if the sauce is too thick.

LATIFA BENNANI-SMIRES
MOROCCAN COOKING

Beans, lentils and peas need boiling to destroy toxins (box, page 15).

Sautéed Lentils
Lentejas Salteadas

To serve 4

¾ cup	dried lentils	175 ml.
3 tbsp.	butter	45 ml.
¼ cup	chopped fresh parsley	50 ml.
3 tbsp.	strained fresh lemon juice	45 ml.
	pepper	
16	triangular croutons	16
4	eggs, hard-boiled and sliced	4
6 oz.	bacon, fried until crisp	175 g.

Put the lentils in enough salted water to cover them by 2 inches [5 cm.] and cook them for 15 to 20 minutes, or until tender but not mushy. Drain them.

Melt the butter in a heavy frying pan. Add the lentils and cook them until they begin to color. Add the parsley, lemon juice and a pinch of pepper; continue cooking, stirring frequently, until the lentils are dry—about five minutes. Serve garnished with the croutons and the egg and the bacon.

MANUEL M. PUGA Y PARGA
LA COCINA PRÁCTICA

Lentils with Hard-boiled Eggs

To serve 4 to 6

2½ cups	dried lentils	625 ml.
4 tbsp.	beef drippings	60 ml.
1	onion, chopped	1
2	carrots, sliced	2
1	celery rib, diced	1
1	garlic clove, finely chopped	1
about 2 cups	beef stock (recipe, page 166)	about ½ liter
½ lb.	slab bacon or ham bone	¼ kg.
	pepper	
6	eggs, hard-boiled	6

Melt the drippings in a heavy pan set over medium heat. Add the chopped onion and sauté for five to 10 minutes, until golden. Add the carrots and celery and stir them around for a few minutes.

Add the lentils, and stir them until they are well coated with fat. Add the garlic and pour on the stock—it should almost cover the lentils. Put in the bacon or ham bone, season with black pepper, and boil the mixture gently for 15 to 20 minutes, or until the lentils are soft. The liquid should have almost boiled away. Pour the lentils and remaining liquid into a serving dish. Halve the eggs and lay them on top. Throw away the bacon unless it is a nice piece, in which case cut it in slices and lay them on top with the eggs.

ARABELLA BOXER
NATURE'S HARVEST

Lentils and Oxtail with Cream
Sahne-Linsen mit Ochsenschwanz

To lard an onion, pierce a hole through the center and draw a strip of lard through it with a larding needle. Then push in a sliver of garlic.

To serve 4 to 6

2 cups	dried lentils	½ liter
2 quarts	meat stock (recipe, page 166)	2 liters
2	large onions, 1 larded, 1 sliced	2
3 tbsp.	chopped cooked ham	45 ml.
2	medium-sized potatoes, cut into chunks	2
1	oxtail, cut between the joints into 2-inch [5-cm.] pieces	1
2 tbsp.	oil	30 ml.
1	garlic clove, crushed	1
½	bunch celery, diced	½
1	leek, cut into strips	1
2	carrots, sliced	2
2 tbsp.	puréed tomato (recipe, page 166)	30 ml.
4 cups	dry red wine	1 liter
1 cup	sour cream	¼ liter
	salt and pepper	

Put the lentils in the meat stock with the whole onion, ham and potatoes, and cook them until the lentils are tender—15 to 20 minutes.

In another pan, fry the oxtail in the oil over high heat, turning the pieces frequently to brown them evenly. Cook for 10 minutes, then add the garlic, celery, leek, carrots, sliced onion and puréed tomato. Reduce the heat and cook for another five minutes, stirring frequently.

Pour the wine over the oxtail, cover the pan, and simmer the mixture gently for about two hours, or until the meat is tender. Remove the oxtail, bone it while it is still warm and cut the flesh into cubes.

Add the meat, vegetables and stock to the lentils. Pour in the sour cream and bring the liquid to a boil. Season with salt and pepper just before serving.

HANS KARL ADAM
DAS KOCHBUCH AUS SCHWABEN

Lentils with Dried Apricots

Linsen mit Backaprikosen

To serve 4

1 cup	dried lentils	¼ liter
½ cup	dried apricots, soaked for 15 minutes in warm water and drained	125 ml.
1	large onion, finely chopped	1
3 tbsp.	butter	45 ml.
	salt and pepper	
4	walnuts, chopped	4
2 tbsp.	chopped fresh parsley or coriander leaves	30 ml.

Place the lentils in a large saucepan, cover them with salted water, cover the pan and cook the lentils for 15 to 20 minutes, or until they are tender. Drain them.

Fry the apricots and onion in the butter over low heat until the onion begins to soften. Season the mixture with salt and pepper; add it, and the walnuts, to the cooked lentils.

Place the saucepan over very low heat for about 10 minutes so the lentils heat through but do not dry out. Put the lentils into a serving dish, sprinkle them with the chopped parsley or coriander leaves, and serve.

KULINARISCHE GERICHTE

Lentil and Sausage Casseroles

To serve 4

1¼ cups	dried lentils	300 ml.
¾ lb.	sweet Italian sausage, sliced	350 g.
¼ lb.	hot Italian sausage, sliced	125 g.
2	onions, chopped	2
1	small garlic clove, finely chopped	1
½	bay leaf	½
½ tsp.	dried oregano leaves	2 ml.
3 cups	beef or chicken stock (recipes, page 166)	¾ liter
2 tbsp.	wine vinegar	30 ml.
3	plum tomatoes, peeled	3
1	small zucchini, sliced	1
	salt and freshly ground pepper	
	chopped fresh fennel greens or parsley	

In a heavy skillet, sauté the sausage slices until done — about 15 minutes. Remove them with a slotted spoon and set them aside on paper towels to drain. In the sausage fat remaining in the skillet, sauté the onions and garlic until the onions are tender but not browned. Add the bay leaf, oregano, stock, vinegar and lentils.

Bring the mixture to a boil, cover the skillet, and simmer for 15 minutes. Add the tomatoes and zucchini, and season with salt and pepper. Cover and continue simmering until the lentils and zucchini are tender but still retain their shapes — about 15 minutes.

Return the sausage slices to the skillet and reheat the mixture. Serve in individual casseroles and sprinkle with the chopped fennel or parsley.

JEAN HEWITT
THE NEW YORK TIMES WEEKEND COOKBOOK

Lentils in Red Wine

Lentilles au Vin Rouge

No bistro in France would make up a winter menu without lentils somewhere on the list. Chunks of bacon are usually added, but you will find that they are not at all necessary. A few extra herbs and spices more than make up the difference. *Lentilles au Vin Rouge* makes an excellent accompaniment to game, guinea hen or Rock Cornish hen. The dish is commonly served hot but is also excellent served cold.

To serve 6

1½ cups	dried lentils	375 ml.
6 cups	dry red wine	1½ liters
3	medium-sized carrots, quartered lengthwise, then thinly sliced crosswise	3
3	medium-sized onions, finely chopped, plus 1 small onion stuck with 3 cloves	3
3	garlic cloves, crushed	3
1 tsp.	ground coriander seeds	5 ml.
1 tbsp.	salt	15 ml.
1 tsp.	pepper	5 ml.
	bouquet garni of 6 sprigs parsley tied around 2 bay leaves	

Place the lentils in a large, heavy pot. Add the red wine, carrots, onions, garlic, seasonings and bouquet garni.

Cover the pot and bring the liquid to a simmer. Cook the lentils for about one hour, or until they are soft but not mushy. Stir rather frequently during the cooking, mixing the lentils so that those no longer immersed in the liquid exchange places with those that are. If you feel that the liquid is evaporating too rapidly, add a little more wine or some water. Even when the lentils are completely cooked, there should be a little bit of liquid remaining. Before serving the lentils, remove the bouquet garni and the small whole onion stuck with cloves.

CAROL CUTLER
HAUTE CUISINE FOR YOUR HEART'S DELIGHT

Beans, lentils and peas need boiling to destroy toxins (box, page 15).

Lentils with Chard and Lemon

'Adas bi Haamud

The unused chard stems may be saved and boiled to serve as a salad with sesame oil.

To serve 6		
1½ cups	dried lentils	375 ml.
2½ lb.	Swiss chard, stems removed but a few stems reserved and chopped with the leaves	1¼ kg.
¾ cup	chopped onion	175 ml.
¾ cup	olive oil	175 ml.
5	garlic cloves	5
1½ tsp.	salt	7 ml.
⅓ cup	chopped fresh coriander leaves or 1 celery rib, chopped	75 ml.
¾ cup	strained fresh lemon juice	175 ml.
1 tsp.	flour	5 ml.
	pepper	

Cook the lentils in enough water to cover them for about 15 to 20 minutes, or until tender. Add the chard and cook for another 10 minutes.

Meanwhile fry the onion in the olive oil. Crush the garlic with the salt and add to the fried onion. Stir the onion and coriander or celery into the chard-and-lentil mixture. Mix the lemon juice with the flour and stir into the soup.

Let the soup simmer until thick. Add pepper and taste for salt. Serve hot.

MARIE KARAM KHAYAT AND MARGARET CLARK KEATINGE
FOOD FROM THE ARAB WORLD

Lucknow Sour Lentils

Lakhnawi Khatti Dal

The technique of making this dish is demonstrated on pages 18-19. Tamarind pulp, the compressed pulp from the peeled and pitted pod of the same name, and mango powder, a dried, ground form of its namesake, are both available in Indian food markets. Indian vegetable shortening (vanaspati ghee) is a product of highly saturated oils such as coconut, cotton-seed, rapeseed and palm; it has a light lemon color, grainy texture and faint nutty, lemony aroma.

To serve 6		
1½ cups	dried pink lentils	375 ml.
½ tsp.	turmeric	2 ml.
1 tbsp.	peeled and grated fresh ginger	15 ml.
1	1-inch [2½-cm.] ball tamarind pulp or 1 tsp. [5 ml.] mango powder or 1 tbsp. [15 ml.] lemon juice	1
2 tsp.	coarse salt	10 ml.
Spice-perfumed butter		
5 tbsp.	Indian vegetable shortening or light vegetable oil	75 ml.
½ tsp.	cumin seeds	2 ml.
¼ to ½ tsp.	cayenne pepper	1 to 2 ml.
1 tbsp.	finely chopped garlic	15 ml.

Put the lentils in a deep saucepan along with the turmeric, ginger and 5 cups [1¼ liters] of water; bring the water to a boil, stirring often, as the lentils have a tendency to settle. Reduce the heat to medium low and simmer, partially covered, for 25 minutes, stirring now and then.

While the lentils are cooking, soak the tamarind pulp—if you are using it—in 1 cup [¼ liter] of boiling water for 15 minutes. Mash the pulp with the back of a spoon or with your fingers. Strain the liquid into a bowl, squeezing out as much juice as possible from the pulp. Discard the stringy residue.

When the lentils have simmered for 25 minutes, add the tamarind juice. If you are using mango powder or lemon juice in place of the tamarind juice, do not add it yet.

Cook the lentils for 15 minutes more. Turn off the heat and, with a wire whisk or wooden spoon, beat the lentils to a smooth purée. Measure the purée and, if necessary, add enough water to make 6 cups [1½ liters]. If you are using mango powder or lemon juice, stir it in now. Add the salt, return the purée to the heat, and simmer it slowly until it is piping hot—about five minutes.

Taste the purée for salt, and put it in a serving bowl while you make the spice-perfumed butter: Put the shortening over medium-high heat in a small skillet. When it is very hot, add the cumin seeds and fry for about 10 seconds.

Remove the pan from the heat and add the cayenne pepper and the garlic; stir rapidly for 10 seconds, or until the garlic loses its raw smell and begins to color. Do not let the garlic brown. Pour the spice-perfumed butter over the lentil purée. Stir once or twice—just enough to lace the purée with ribbons of the spice-perfumed butter—and serve immediately in small bowls.

JULIE SAHNI
CLASSIC INDIAN COOKING

Curried Red Lentils

Masur Dal

To prepare freshly grated coconut, first puncture the three eyes at one end of the shell, drain out the coconut liquid and split the shell in half. Pry the meat from the shell with a small knife, and pare off and discard the papery brown skin; then grate the meat. In a tightly covered container, the grated coconut can be safely kept in the refrigerator for up to two days.

To serve 4

1 cup	dried red lentils	¼ liter
1 tsp.	salt	5 ml.
2 tbsp.	*ghee*	30 ml.
2	onions, chopped	2

Masala paste

1 tsp.	cumin seeds	5 ml.
1 tsp.	poppy seeds	5 ml.
1 tsp.	paprika or chili powder	5 ml.
1 tsp.	turmeric	5 ml.
2 tsp.	coriander seeds	10 ml.
6	whole cloves	6
2-inch	cinnamon stick	5-cm.
4	green cardamom pods	4
1 cup	freshly grated coconut	¼ liter
4	peppercorns	4
4	garlic cloves	4

Cover the lentils with water, add the salt and bring to a boil. Cook over medium heat for 30 minutes, or until the lentils are soft and have absorbed the water. Meanwhile, pound the ingredients for the *masala* paste with a mortar and pestle, or grind them to a paste in a blender or food processor.

Heat the *ghee* and fry the onions in it until they are golden. Add the *masala* paste and fry for a few minutes to blend the flavors. Stir this mixture into the cooked lentils. Serve hot with rice.

JACK SANTA MARIA
INDIAN VEGETARIAN COOKERY

Peas

Meatballs with Zucchini and Chick-peas

Karyaprak

To serve 6

1¼ cups	dried chick-peas, soaked overnight and drained	300 ml.
1 lb.	beef chuck, finely chopped	½ kg.
3	slices firm-textured white bread, crusts removed, soaked in water and squeezed dry	3
4	garlic cloves, finely chopped	4
2 tbsp.	chopped fresh parsley	30 ml.
	salt and pepper	
	quatre épices	
½ tsp.	freshly grated nutmeg	2 ml.
2	eggs	2
about ¼ cup	flour	about 50 ml.
⅔ cup	oil	150 ml.
3	onions, chopped	3
4	tomatoes, peeled, seeded and chopped	4
2 lb.	zucchini, cut into thick rounds	1 kg.

Parboil the chick-peas in enough water to cover them for about one hour. Meanwhile, combine the beef, bread crumbs, garlic and parsley. Season the mixture with salt and pepper, the *quatre épices* and the nutmeg. Add one of the eggs and mix well. Form the mixture into 12 small meatballs.

Beat the remaining egg. Roll the meatballs in the flour, then dip them in the beaten egg. Heat the oil and fry the meatballs in it for 15 minutes, turning them frequently to brown them evenly. Take the balls from the pan and, in the same oil, gently cook the onions and tomatoes for 20 minutes; season them with salt and pepper and add 2 cups [½ liter] of hot water. Simmer the mixture for a few minutes, then force it through a sieve. Set the resulting sauce aside.

Drain the chick-peas and lay them in the bottom of an earthenware dish; cover them with the zucchini and put the meatballs on top. Pour the sauce over all, cover the dish with a piece of foil in which several slits have been made, and cook in a preheated 350° F. [180° C.] oven for two hours.

IRÈNE AND LUCIENNE KARSENTY
LA CUISINE PIED-NOIR

Beans, lentils and peas need boiling to destroy toxins (box, page 15).

Stew with Chick-peas
T'fina aux Pois Chiches

To serve 6

1½ cups	dried chick-peas, soaked overnight and drained	375 ml.
2 to 2½ lb.	beef brisket	1 to 1¼ kg.
1	calf's foot	1
10 to 12	potatoes, soaked in salted water for several hours	10 to 12
6	eggs	6
1	garlic bulb	1
1 tsp.	honey	5 ml.
1 tsp.	paprika	5 ml.
1 tbsp.	olive oil	15 ml.
	salt and pepper	

Meatball

10 oz.	ground beef	300 g.
3	slices firm-textured white bread, crusts removed, soaked in water and squeezed dry	3
	freshly grated nutmeg	
	ground mace	
	quatre épices	
	salt and pepper	
1 tbsp.	chopped fresh parsley	15 ml.
1	garlic clove, chopped	1
1	egg	1

To make the meatball, mix together the ground beef, bread, nutmeg, mace, *quatre épices*, salt, pepper, parsley and garlic, and bind it with the egg. Put it aside.

In a thick-bottomed casserole, put the brisket, calf's foot, chick-peas, potatoes and the six eggs, still in their shells. Add the whole garlic bulb with the honey, paprika, olive oil, salt and pepper. Pour on 2 quarts [2 liters] of water, bring to a boil, add the meatball, cover, and simmer for four hours.

Uncover the casserole and place it in a preheated 250° F. [130° C.] oven. Cook the stew for one hour or more, until very little liquid remains in the pot.

Shell the eggs before serving. Put the chick-peas, potatoes and eggs in a deep serving dish; arrange the meats on a separate platter.

IRÈNE AND LUCIENNE KARSENTY
LA CUISINE PIED-NOIR

Meat and Chick-pea Stew
Cocido Madrileno

To serve 4

1 cup	dried chick-peas, soaked overnight and drained	¼ kg.
1 lb.	beef round or chuck, in 1 piece	½ kg.
¼ lb.	cured ham, in 1 piece	125 g.
1 lb.	stewing chicken	½ kg.
1 lb.	veal shank, cut into 2 pieces	½ kg.
	salt	
3	medium-sized carrots	3
3	medium-sized onions	3
1 or 2	turnips	1 or 2
4	small potatoes	4
2	small *chorizos* (about 6 oz. [175 g.])	2
2	small blood sausages (about 6 oz. [175 g.])	2
3	medium-sized leeks	3
1½ lb.	cabbage, cut into large pieces	¾ kg.
¾ cup	puréed tomato *(recipe, page 166)*	175 ml.

Put the beef, ham, chicken and bones into a large pot. Pour in about 4 quarts [4 liters] of water. Cover the pot and bring the water to a boil—the pot should be covered throughout the entire cooking of this dish. Tie the chick-peas in a small net or cheesecloth bag, and toss them into the pot. Reduce heat, cover, and cook until the chick-peas are tender—one and one half to two hours. Season with salt and add the carrots, onions and turnips. Continue to boil for about another hour, then add the potatoes, *chorizos,* blood sausages and leeks.

Boil the *cocido* for about one hour more to complete its cooking; if the chick-peas begin getting crumbly, remove them until about 10 minutes before serving.

Meanwhile, boil the cabbage in salted water in a separate pan until it is tender. Before serving the *cocido,* add the cabbage with its cooking liquid.

To present this dish, open with a bowl of the broth for each person. Then heat up the puréed tomato and pour it into a saucer or gravy bowl; cut the *cocido* ingredients into four helpings. Present the vegetables on one platter, the meats and chick-peas on another—both without broth—and the puréed tomato on the side.

DONN E. POHREN
ADVENTURES IN TASTE

Falafel

Falafel is a dish common to Egypt and Israel, and it is also eaten throughout the Near East. Street vendors sell these fried chick-pea balls, which are stuffed inside pocket-shaped Arab pita bread together with lettuce and tahini (sesame-seed paste). Falafel is also served as an appetizer on a small plate of hummus (ground chick-pea paste), garnished with quartered hard-boiled eggs and sprinkled with chopped parsley, cayenne pepper, tahini and a few drops of rich green olive oil.

To serve 4

2½ cups	dried chick-peas, soaked overnight and drained	625 ml.
1 tsp.	ground coriander seeds	5 ml.
1	garlic clove, chopped	1
1 tsp.	ground cumin	5 ml.
½ tsp.	cayenne pepper	2 ml.
	salt	
¼ cup	flour	50 ml.
	oil for deep frying	

Grind the chick-peas fine in a blender or food processor, and mix them with the coriander, garlic, cumin, cayenne pepper and salt. Add the flour and mix thoroughly. From the resulting dough, make small balls about 1¼ inches [3 cm.] in diameter. Pour oil into a pan and heat it to 375° F. [190° C.]. A few at a time, deep fry the balls for two to three minutes until they are golden.

NAOMI & SHIMON TZABAR
YEMENITE & SABRA COOKERY

Friday Chick-peas

Les Pois Chiches du Vendredi

Traditionally, chick-peas were served on a Friday with a *brandade* of salt cod and puréed potatoes.

To serve 4 to 6

2 cups	dried chick-peas, soaked overnight and drained	½ liter
2	large carrots	2
1	large onion, stuck with 1 whole clove	1
3	garlic cloves	3
¼ cup	olive oil	50 ml.
	salt	
1 tbsp.	white wine vinegar	15 ml.
	pepper	

Rinse the chick-peas thoroughly and place them in a large, deep saucepan. Pour in enough water to cover the peas well.

The pan should be big enough for the water to reach no more than two thirds of the way up the sides. Add the carrots, onion, garlic and 3 tablespoons [45 ml.] of the olive oil. Bring quickly to a boil and then reduce to a simmer. Cover the pan and cook the peas for two hours, taking care that the water does not boil over or the correct combination of water, oil and vegetable juices will be lost. Add a little salt, cover again and cook for another hour.

Drain the chick-peas. Reserve the cooking liquid and discard the carrot, onion and garlic. Serve the chick-peas hot with a little cooking liquid, a dash of wine vinegar, the remaining olive oil and a little ground pepper.

ALBIN MARTY
FOURMIGUETTO

Chick-pea Dip

Cecilina

To toast sesame seeds, place them on a baking sheet and set the sheet in a preheated 400° F. [200° C.] oven for about five minutes, or until the seeds are lightly browned. Shake the sheet periodically to ensure that the seeds brown evenly. Corn tortillas may be purchased at Latin American groceries and many supermarkets. To prepare them for chips, first cut them into pieces. Heat ½ to 1 inch [1 to 2½ cm.] of oil in a skillet until a drop of water tossed in hisses. Fry the tortilla pieces—do not crowd them—for a minute on each side; they should be golden with crisp edges. Drain the chips on paper towels.

To serve 6 to 8

1 cup	dried chick-peas, soaked overnight and drained	¼ liter
3 tbsp.	butter	45 ml.
1 cup	finely chopped onion	¼ liter
2 tbsp.	finely chopped parsley	30 ml.
3 tbsp.	coarsely chopped pine nuts	45 ml.
½ tsp.	ground oregano	2 ml.
	Tabasco sauce	
2 tbsp.	sesame seeds, toasted	30 ml.
	crisp-fried corn tortilla chips	

Cook the chick-peas in boiling water for about one and one half hours, or until tender. Drain the peas thoroughly. Purée them in a blender or food mill, then pass them through a drum sieve to remove the skins.

Melt the butter in a small skillet and cook the onion over medium heat until it is transparent—about five minutes. Blend the onion into the pea purée along with the parsley, pine nuts, oregano and a dash of Tabasco sauce. Turn the purée into a serving bowl, cover it and chill.

At serving time, sprinkle the purée with the sesame seeds. Serve as an appetizer dip with the tortilla chips.

ALEX D. HAWKES
A WORLD OF VEGETABLE COOKERY

Beans, lentils and peas need boiling to destroy toxins (box, page 15).

Chick-peas with Green Peppers and Chilies

The volatile oils in hot chilies may make your skin sting and your eyes burn; after handling chilies, avoid touching your face and wash your hands promptly.

If hot green chilies are not available, use ½ teaspoon [2 ml.] of Tabasco sauce, adding it with the lemon juice.

To serve 4

1 cup	dried chick-peas, soaked overnight and drained	¼ liter
3 tbsp.	olive oil	45 ml.
2	medium-sized green peppers, seeded, deribbed and diced	2
4	fresh hot green chilies, stemmed, seeded and finely chopped	4
1	large garlic clove, finely chopped	1
3 tbsp.	strained fresh lemon juice	45 ml.

Place the chick-peas in a large pot, cover them with water, cover the pot and cook the chick-peas for three hours, or until tender. Drain the chick-peas.

Heat the oil in a skillet over low heat. Stir in the green peppers, chilies and garlic; cook for five minutes.

Add the chick-peas, and cook them until they are just showing signs of browning. Pour in the lemon juice and let it bubble. Serve the chick-peas as soon as you can.

GAIL DUFF
GAIL DUFF'S VEGETARIAN COOKBOOK

Chick-peas Majorcan-Style
Garabanzos

To serve 6 to 8

2½ cups	dried chick-peas, soaked overnight and drained	625 ml.
1	slice pumpkin (about ½ lb. [¼ kg.]), peeled and seeded	1
3	garlic cloves, chopped	3
1	medium-sized tomato, peeled, seeded and chopped	1
1	egg, hard-boiled and finely chopped	1
	salt and pepper	
¼ cup	finely chopped fresh parsley	50 ml.

Place the chick-peas in a pan with 2½ quarts [2½ liters] of cold water; add the pumpkin and cook, covered, over low heat for two to three hours, or until the chick-peas are soft.

When the chick-peas are cooked, drain off the liquid, remove the pumpkin and purée it through a sieve. In a glazed earthenware pot, fry the garlic in the olive oil. When the garlic is lightly browned, add the tomato. After about five minutes, when the tomato is soft, add the puréed pumpkin and the hard-boiled egg. Season with salt and pepper, add the parsley and stir in the chick-peas. Mix well and serve.

COLOMA ABRINAS VIDAL
COCINA SELECTA MALLORQUINA

Pickled Black-eyed Peas

The cooking of black-eyed peas is explained on pages 16-17.

To serve 8

4 cups	cooked black-eyed peas	1 liter
1 cup	oil	¼ liter
¼ cup	wine vinegar	50 ml.
1 cup	whole garlic cloves	¼ liter
1	medium-sized onion, thickly sliced	1
½ tsp.	salt	2 ml.
	cracked or freshly ground pepper	

Combine all of the ingredients, mix well, and place them in a jar. Cover the jar and place it in the refrigerator. Remove the garlic after one day, but continue to pickle the beans for as long as two weeks.

MARY FAULK KOOCK
THE TEXAS COOKBOOK

Bean-Dough Cakes
Elele

Chopped green peppers or cooked, finely chopped shrimp may be added to this batter before frying.

To serve 4

2 cups	dried black-eyed peas, soaked overnight and drained	½ liter
1	large onion	1
1	large tomato, peeled	1
2 tbsp.	flour (optional)	30 ml.
1 tsp.	salt	5 ml.
½ tsp.	pepper	2 ml.
1 tsp.	ground ginger	5 ml.
1	egg, lightly beaten (optional)	1
	oil for deep frying	

Rub the soaked peas with your finger tips to loosen and remove the husks; then wash the peas in a bowl of water.

Repeat this several times until the husks and all of the black eyes have been removed and discarded.

In a blender, grind the peas until they are very smooth, then add the vegetables and ½ cup [125 ml.] of water. Grind all together—adding the flour, if used—until the ingredients are well blended and a smooth batter is obtained. Pour the batter into a bowl and set it aside, covered, for about 30 minutes. Stir in the salt, pepper, ginger and, if used, the egg.

Heat oil in a heavy saucepan or deep fryer until it registers 375° F. [190° C.] on a deep-frying thermometer. Drop spoonfuls of batter into the oil, and fry the cakes until all sides are golden brown. (For smoother shapes, dip the spoon into water each time before spooning out the batter.)

DINAH AMELEY AYENSU
THE ART OF WEST AFRICAN COOKING

Bohemian Peas

Boehmische Erbsen

To serve 6

2 cups	dried green or yellow peas, soaked overnight and drained	½ liter
1	sprig thyme	1
1	bay leaf	1
3	whole cloves	3
4	peppercorns	4
2	small onions, thinly sliced	2
1	carrot	1
10 tbsp.	butter	150 ml.
3 tbsp.	dry bread crumbs, 2 tbsp. [30 ml.] sautéed in 2 tsp. [10 ml.] butter	45 ml.
	salt and pepper	

Wrap the thyme, bay leaf, cloves, peppercorns and a few slices of onion in a square of cheesecloth. Put the wrapped seasonings, the peas and the carrot in a pot, and cover them with water—about 6 cups [1½ liters]. Bring the water to a boil. Skim the pea skins from the pot as they rise to the surface. When the peas are soft, after one to one and one half hours, strain them, keeping the cooking liquid for making soup. Discard the carrot and the wrapped seasonings.

Melt 3 tablespoons [45 ml.] of the butter in a skillet, and fry the remaining onion slices in it until they are golden. Set the onions aside.

Coat a 6-cup [1½-liter] baking dish liberally with the remaining butter. Sprinkle the dish with the dry bread crumbs. Season the drained peas with salt and pepper, then pile them into the dish. Top the peas with the fried onions, then cover with the sautéed bread crumbs. Cook in a preheated 375° F. [190° C.] oven for about 15 minutes to crisp the top. Serve immediately, directly from the baking dish.

ROSL PHILPOT
VIENNESE COOKERY

Purée of Dried Yellow Peas

Erbsenpüree

To serve 4 to 6

2¾ cups	dried yellow peas, soaked overnight and drained	675 ml.
1	small bunch fresh parsley, tied	1
¼ lb.	bacon rind	125 g.
	salt and pepper	
2 tbsp.	butter	30 ml.

Put the peas in a saucepan together with the parsley (including some parsley root if possible), bacon rind, salt and pepper, and enough water to cover them. Cover the pan and simmer until the peas are quite soft—one and one half to two hours. Drain the peas, discard the parsley and rind, and then force the peas through a sieve. Add a little of the water in which they were cooked to obtain a stiff but creamy consistency. Beat the butter into the hot purée and serve.

GRETEL BEER
AUSTRIAN COOKING AND BAKING

Yellow-Pea Purée

Erbsenpüree

Served with sauerkraut and pork knuckles, this is a traditional Berlin dish and a great favorite in northern Germany generally. In place of the meat stock, you may cook the peas with smoked or salted pork knuckles or ribs, cut into neat pieces and added to the peas with 2 cups [½ liter] of water. The meat may then be served with the puréed peas.

To serve 6

2 cups	dried yellow peas, soaked overnight and drained	½ liter
2 cups	meat stock (recipe, page 166)	½ liter
	salt and pepper	
	dried marjoram leaves	
4 tbsp.	butter	60 ml.
2	onions, chopped	2

Place the peas in a pan with 5 cups [1¼ liters] of water. Cook until most of the water has been absorbed and the peas are swollen but still whole. Add the stock and continue to cook until the peas are tender.

Drain the peas; pass them through a sieve, and season the purée with salt, pepper and marjoram. Stir in half of the butter. Keep the purée warm. Brown the onions in the remaining butter and pour them over the puréed peas.

GRETE WILLINSKY
KOCHBUCH DER BÜCHERGILDE

Beans, lentils and peas need boiling to destroy toxins (box, page 15).

Pigeon-Pea Casserole

The salt-cured beef specified in this recipe is seasoned, sun-dried meat, sold as tasajo in Latin American markets. For a milder casserole, diced fresh beef may be substituted.

To serve 6 to 8

2 cups	dried pigeon peas	½ liter
½ lb.	salt-cured beef, diced, soaked for 20 minutes and drained	¼ kg.
1	medium-sized tomato, chopped	1
2	garlic cloves	2
1	large onion, sliced	1
4 tbsp.	butter	60 ml.
1	medium-sized potato, diced	1
¾ cup	peeled, seeded and diced pumpkin	175 ml.
1	carrot, sliced	1
1 tsp.	whole thyme leaves	5 ml.
1 tsp.	salt (optional)	5 ml.

Brown the meat in its own fat and drain it. Brown the tomato, garlic and onion in the butter in the bottom of a large, heavy casserole. Add the beef, all of the remaining ingredients and 4 cups [1 liter] of water. Cover the casserole and cook, adding more water if necessary, for about one and one half hours, or until the meat and peas are tender.

CONNIE AND ARNOLD KROCHMAL
CARIBBEAN COOKING

Pease Pudding

To serve 6

2 cups	dried split peas, soaked for 1 hour and drained, or dried whole peas, soaked overnight and drained	½ liter
4 tbsp.	butter, cut into small pieces	60 ml.
1	large egg	1
	salt and pepper	

Put the peas in a pan, cover them with plenty of fresh water and cook until tender. Split peas will take from 45 to 60 minutes; whole dried peas will need at least two hours. Drain off the liquid—keep it for soup—and put the beans through a food processor briefly to make a purée that is not too smooth. Mix the butter into the purée, then add the egg and season well with salt and pepper. Put the pudding mixture into a buttered 5-cup [1¼-liter] pudding basin or heat-proof dish, cover with foil or a cloth, and steam for one hour in a pan of boiling water. Turn out the pudding and serve it.

JANE GRIGSON
ENGLISH FOOD

Spiced Yellow Split Peas

Amti

The volatile oils in hot chilies may make your skin sting and your eyes burn; after handling chilies, avoid touching your face and wash your hands promptly.

To serve 4

1 cup	dried yellow split peas, soaked for 1 hour and drained	¼ liter
2 tbsp.	ghee	30 ml.
½ tsp.	mustard seeds	2 ml.
3	fresh hot green chilies, stemmed, seeded and chopped (optional)	3
1	onion, finely chopped	1
½ tsp.	turmeric	2 ml.
½ tsp.	paprika or chili powder	2 ml.
12	peppercorns	12
1-inch	cinnamon stick, broken into pieces	2½-cm.
4	whole cloves	4
2	green cardamom pods	2
1 tbsp.	strained fresh lemon juice	15 ml.
1 tsp.	salt	5 ml.

Boil the split peas in 2½ cups [625 ml.] of water for 40 minutes, or until they are soft and all of the water is absorbed.

In a frying pan, heat the *ghee* and fry the mustard seeds until they sputter. Add the chilies and onion and fry until golden. Stir in the turmeric and the paprika or chili powder. Crush the peppercorns, cinnamon, cloves and cardamom in a mortar and add these to the frying pan. Fry for two minutes. Stir this mixture into the split peas. Add the lemon juice and salt, mix well and simmer for a few minutes. Serve hot.

JACK SANTA MARIA
INDIAN VEGETARIAN COOKERY

Curried Yellow Split Peas

Channa Dhal

The original version of this recipe calls for chana dal, a dried split pea similar to American yellow split peas, and sometimes obtainable at Indian food markets. To prepare freshly grated coconut, see the editor's note for Curried Red Lentils, page 115.

The volatile oils in hot chilies may make your skin sting

and your eyes burn; after handling chilies, avoid touching your face and wash your hands promptly.

	To serve 6	
1¼ cups	dried yellow split peas, soaked for 1 hour and drained	300 ml.
	salt	
1 tsp.	turmeric	5 ml.
2 tbsp.	*ghee*	30 ml.
2	onions, chopped	2
1 tbsp.	freshly grated coconut	15 ml.
3	dried hot chilies, each cut into 2 or 3 pieces	3
1 tsp.	ground cumin	5 ml.
2 or 3	bay leaves	2 or 3
2 tsp.	peeled and grated fresh ginger	10 ml.
1 tsp.	*garam masala*	5 ml.
2 or 3	fresh hot chilies, thickly sliced	2 or 3
1 tsp.	sugar	5 ml.

Cook the peas in 4 cups [1 liter] of salted water with the turmeric. When the peas are soft and cooked—about 35 minutes—remove them from the heat.

Heat the *ghee* in a saucepan, and brown the onion and coconut with the dried chilies, cumin, bay leaves and ginger. Add the *garam masala* and fry for two to three minutes.

Add the cooked peas plus the fresh chilies and sugar to the fried mixture, and boil for a few minutes.

P. MAJUMDER
COOK INDIAN

Split Pea Fritters

Opiekanki Grochu

	To serve 6	
2 cups	dried green or yellow split peas	½ liter
1	onion, finely chopped	1
4 tbsp.	butter	60 ml.
about ½ cup	dry bread crumbs	about 125 ml.
1 tbsp.	finely cut mixed fresh dill and parsley	15 ml.
	salt and pepper	
1	egg, lightly beaten	1

Put the peas in enough water to cover them, and cook them for about 40 minutes until tender. Drain the beans and purée them without diluting them with liquid. Brown the onion in

1 tablespoon [15 ml.] of butter and add to the purée, together with ¼ cup [50 ml.] of the bread crumbs and the dill and parsley. Season with salt and pepper, then add half of the beaten egg to make a thick batter.

Shape the batter into patties or fritters, dip them in the remaining egg, roll them in the remaining bread crumbs, and fry them in the remaining 3 tablespoons [45 ml.] of butter to a golden brown on all sides.

MARJA OCHOROWICZ-MONATOWA
POLISH COOKING

Rice and Wild Rice

Jaipur Rice Salad

The techniques of cooking rice are explained on pages 22-23.

	To serve 6	
3 cups	cooked white rice, cold	¾ liter
2	tomatoes, peeled, seeded and cut into small cubes	2
6	radishes, thinly sliced	6
½ cup	finely chopped onion	125 ml.
½ cup	finely chopped celery heart	125 ml.
4	small beets, cooked, peeled, cooled and cut into thin strips	4
¼ cup	sliced green olives	50 ml.
	salt and freshly ground pepper	
½ tsp.	curry powder	2 ml.
1 tsp.	Dijon mustard	5 ml.
2 tbsp.	wine vinegar	30 ml.
6 tbsp.	olive oil	90 ml.

Combine the cold rice with the tomatoes, radishes, onion, celery, beets, olives, and salt and pepper to taste. Combine the curry powder, mustard and vinegar; sprinkle this mixture over the salad. Toss the salad lightly and sprinkle with the oil. Toss again and serve.

RENÉ VERDON
THE WHITE HOUSE CHEF COOKBOOK

Callaloo and Rice

Callaloo, a Caribbean leafy vegetable, is sometimes obtainable at Latin American markets. If it is unavailable, substitute fresh spinach or Swiss chard.

To serve 6

2 cups	unprocessed white rice	½ liter
½ lb.	callaloo	¼ kg.
1	small onion, sliced	1
1	garlic clove, chopped	1
1	lemon, sliced, including peel	1
1	large tomato, quartered	1
2½ oz.	cooked ham, diced (about ½ cup [125 ml.])	75 g.

Combine all of the ingredients with 4½ cups [1,125 ml.] of water and bring to a boil. Cover the pan and simmer gently for 30 to 40 minutes, or until the rice is tender.

CONNIE AND ARNOLD KROCHMAL
CARIBBEAN COOKING

Chilean Eel Curry

Chilenischer Aal-Curry

Eel should be bought live; to prepare one, first stun it by knocking its head against a work surface, then pierce its head with a small knife to kill it. Slit the skin all around the head, loosen a tab of skin with pliers, and then, using the tab, peel back the skin from the head as if pulling off a stocking. Decapitate the eel about 3 inches [7½ cm.] behind the head. Discard the head, which contains the viscera.

To serve 6

2 cups	unprocessed long-grain white rice	½ liter
4 cups	dry white wine	1 liter
2	bay leaves	2
3	whole cloves	3
6	juniper berries	6
1 lb.	eel	½ kg.
2 tbsp.	butter	30 ml.
	salt	
1 tbsp.	curry powder	15 ml.
1 tbsp.	finely cut fresh dill	15 ml.

Bring the white wine to a boil with the bay leaves, cloves and juniper berries. Simmer for five minutes. Put in the eel and poach it for about 10 minutes; remove it and set it aside, reserving the liquid. Fry the rice briskly in the butter. Strain the eel poaching liquid and add it to the rice. Simmer for 10 minutes, or until the liquid is mostly absorbed. Season

with salt and the curry powder. Remove the bones from the eel and cut it into pieces 1 inch [2½ cm.] long. Add it and the dill to the rice. Cover and cook for 10 minutes over low heat.

LILO AUREDEN
WAS MÄNNERN SO GUT SCHMECKT

Rice with Chicken

Arroz con Pollo

To serve 6 to 8

¾ cup	unprocessed long-grain white rice	175 ml.
3 lb.	chicken, cut into 8 serving pieces	1 ½ kg.
	salt and pepper	
about ⅓ cup	flour	about 75 ml.
1 to 2 tbsp.	olive oil	15 to 30 ml.
1 tbsp.	butter	15 ml.
1	medium-sized onion, finely chopped	1
1	garlic clove, finely chopped	1
1	medium-sized green pepper, seeded, deribbed and cut into 1-inch [2½-cm.] squares	1
3	very ripe tomatoes, peeled, seeded and cut into chunks	3
1 ½ cups	chicken stock (recipe, page 166)	375 ml.
½ tsp.	saffron threads	2 ml.
2 tbsp.	dry sherry (optional)	30 ml.
2	bay leaves	2
2 tbsp.	sliced pimiento-stuffed green olives	30 ml.
2 cups	shelled fresh peas	½ liter
¼ cup	finely chopped pimiento	50 ml.

Wash the chicken pieces, dry them thoroughly and sprinkle them with salt and pepper. Dust them lightly with flour. In a large, deep skillet heat 1 tablespoon [15 ml.] of olive oil and the butter to sizzling. Add the chicken in a single layer, skin side down, and brown it lightly. Turn it and brown the other side. Remove the chicken and set it aside.

Put the onion, garlic and green pepper into the skillet. If necessary, add more olive oil. Cook the vegetables lightly and then add the rice. Cook, stirring, a few minutes longer, or until the rice grains are opaque. Add the tomatoes.

Heat the chicken stock, add the saffron and let it steep a few minutes. Pour the stock and sherry, if using, into the skillet and add the bay leaves, olives, green peas and pimiento. Blend the ingredients, add the chicken and arrange it in the skillet. Cover tightly, and cook over low heat until the rice and chicken are tender—35 to 40 minutes.

LILA PERL
RICE, SPICE AND BITTER ORANGES

Country "Dirty" Rice

To serve 6

2 cups	unprocessed long-grain white rice	½ liter
3 tbsp.	flour	45 ml.
½ cup	oil	125 ml.
1	medium-sized onion	1
1½ lb.	chicken livers, finely chopped	¾ kg.
½ cup	chopped celery	125 ml.
½ cup	finely chopped fresh parsley	125 ml.
1 cup	chopped scallions	¼ liter
½ cup	chopped green pepper	125 ml.
1 tsp.	finely chopped garlic	5 ml.
	salt and pepper	
⅛ tsp.	cayenne pepper	½ ml.
4½ cups	meat stock *(recipe, page 166)*	1,125 ml.

In an iron pot, cook the flour in the oil over low heat to make a brown roux. Add the onion and stir until brown. Add the chicken livers, celery, parsley, scallions, green pepper and garlic, and stir. Season with a little salt and pepper and add the cayenne pepper.

Cook the vegetable mixture for five minutes over medium heat; then add ½ cup [125 ml.] of the stock and cook for 15 minutes. Skim any excess oil from the top of the mixture and remove the pan from the heat.

In a separate pan, simmer the rice, covered, in the rest of the stock for 18 to 20 minutes, or until the liquid is absorbed. Fold the cooked rice into the vegetable mixture and correct the seasoning.

Before serving, warm in a preheated 350° F. [180° C.] oven for 10 minutes.

THE JUNIOR LEAGUE OF NEW ORLEANS
THE PLANTATION COOKBOOK

Jefferson Rice

To serve 4 to 6

1 cup	unprocessed long-grain white rice	¼ liter
2 cups	chicken stock *(recipe, page 166)* or water	½ liter
4 tbsp.	butter	60 ml.
1 tsp.	salt	5 ml.
3 oz.	pine nuts (about ½ cup [125 ml.])	90 g.
2 oz.	unsalted pistachios (about ¼ cup [50 ml.])	60 g.
¼ tsp.	ground mace	1 ml.

In a heavy 1- to 2-quart [1- to 2-liter] saucepan, bring the chicken stock or water, 1 tablespoon [15 ml.] of the butter and ½ teaspoon [2 ml.] of the salt to a boil over high heat. Pour in the rice, stir well and reduce the heat to low. Cover the pan tightly and simmer the rice for about 20 minutes, or until it is tender and has absorbed all of the liquid.

Meanwhile, in a heavy skillet, melt the remaining butter over moderate heat. When the foam begins to subside, add the pine nuts and pistachios and, stirring frequently, fry them until they are a delicate golden color. Remove the skillet from the heat.

Transfer the cooked rice to a heated serving bowl and fluff it with a table fork. Then, with a rubber spatula, scrape the contents of the skillet over the rice, and toss the rice and nuts gently together. Sprinkle the pilau with the mace and remaining salt and serve at once.

FOODS OF THE WORLD
AMERICAN COOKING: SOUTHERN STYLE

"Dirty" Rice from Ibiza

Arroz a la Bruta

The techniques of cleaning and preparing squid are demonstrated on page 84. In this recipe, the rice turns black when cooked with the squid's ink—hence the title.

To serve 4

1 cup	unprocessed white rice	¼ liter
about 3 tbsp.	oil	about 45 ml.
1	large onion, chopped	1
1	tomato, peeled, seeded and chopped	1
3 or 4	garlic cloves, chopped	3 or 4
3 tbsp.	chopped fresh parsley	45 ml.
1	large fresh squid (about 1 to 1½ lb. [½ to ¾ kg.]), cleaned, ink sac reserved, body pouch cut into rings	1
	salt and pepper	
	cayenne pepper	
	ground cinnamon	

Heat the oil in a fireproof earthenware pan, and lightly fry the onion, tomato, garlic and parsley. Add the squid rings, and season with salt, pepper, cayenne pepper and cinnamon. Cover the pan and simmer the mixture for about 30 minutes.

Add the rice and let it cook for a few moments, then cover the mixture with about 2 cups [½ liter] of boiling water. Return the water to a boil over high heat.

Meanwhile, break the squid's ink sac into a cup, dilute it with a few spoonfuls of the cooking water, and add it to the rice when it begins to boil. Mix the ink in very well, cover the pan, reduce the heat to medium, and continue cooking the rice—adding more boiling water as necessary—for 15 to 20 minutes, or until a tender, very dry black rice is achieved.

JUAN CASTELLÓ GUASCH
BON PROFIT! EL LIBRO DE LA COCINA IBICENCA

123

Catalan Rice

Arros à la Catalane

To serve 6

2 cups	unprocessed long-grain white rice	½ liter
1 lb.	pork chops	½ kg.
2 tbsp.	olive oil	30 ml.
1	onion, chopped	1
2 to 3 tbsp.	puréed tomato (recipe, page 166)	30 to 45 ml.
5 oz.	chorizo, cut into pieces	150 g.
2 tbsp.	shelled fresh peas	30 ml.
3	garlic cloves, chopped	3
¼ tsp.	powdered saffron	1 ml.
4 to 6 cups	meat stock (recipe, page 166) or water	1 to 1½ liters
	salt and pepper	

In a large saucepan, brown the pork in the olive oil over medium heat for about two minutes. Add the onion and cook for three more minutes until the onion is transparent. Add the puréed tomato, the chorizo, and a little of the stock or water. Cook for 10 minutes, then add the peas and the garlic.

Bring the liquid to a boil, and add the rice, saffron and remaining stock or water. Cook for 15 to 20 minutes, until the rice is soft and the liquid has been absorbed. Season with salt and pepper to taste.

MARIE-THÉRÈSE CARRÉRAS AND GEORGES LAFFORGUE
LES BONNES RECETTES DU PAYS CATALAN

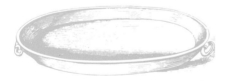

Paella with Chicken, Pork and Seafood

Paella Valenciana

The original version of this recipe calls for Dublin Bay prawns and fresh snails, which are generally unavailable in the United States. Large shrimp may be substituted for the prawns and canned snails for fresh ones. The ingredients in the paella may, in fact, be varied according to taste: A rabbit may be substituted for the chicken, and clams for the mussels; the eel and snails may be omitted entirely.

The technique of cleaning squid is demonstrated on page 84. To clean and prepare eel, see the editor's note for Chilean Eel Curry, page 122. To prepare mussels, scrub the shells and scrape off the beards. Discard any mussels whose shells have opened, then place the mussels in a pan, cover it tightly and cook the mussels over high heat for five minutes; reserve the liquor the mussels exude.

To serve 6 to 8

1¾ cups	unprocessed white rice	425 ml.
¾ cup	olive oil	175 ml.
3 lb.	chicken, cut into 6 or 8 pieces	1½ kg.
½ lb.	lean pork, cut into pieces	¼ kg.
¼ lb.	cooked ham, thinly sliced	125 g.
8	small smoked sausages, pricked in several places	8
2	medium-sized onions, finely chopped	2
4	red peppers, seeded, deribbed and chopped	4
2	artichokes, trimmed and quartered, chokes removed	2
¼ lb.	green beans, trimmed and cut into short lengths	125 g.
2	fresh squid, cleaned and pouches cut into rings	2
1	small eel, skinned, cleaned and cut into 6 or 8 pieces	1
½ lb.	large fresh shrimp	¼ kg.
2	tomatoes, peeled, seeded and chopped	2
2	garlic cloves, chopped	2
6 tbsp.	chopped fresh parsley	90 ml.
	white pepper	
	freshly grated nutmeg	
	salt	
½ tsp.	powdered saffron	2 ml.
1 tsp.	paprika	5 ml.
24	mussels, cooked and shelled, cooking liquid reserved	24
1½ cups	shelled fresh peas, parboiled for 5 minutes and drained	375 ml.
1	bay leaf	1
12	canned snails	12

Place a paella dish, with the oil, over medium heat; when the oil is hot, add the chicken, pork, ham and sausages. Once these have turned a golden color, add the onions, peppers, artichokes and green beans, and continue to cook for a few minutes. Then add the squid and eel. Toss in the shrimp and the tomatoes.

Reduce the heat to low and continue to cook, adding the garlic and parsley and seasoning with white pepper, nutmeg and salt. Add the rice, stir well, and season with a little saffron and paprika. Pour in the mussels' cooking liquid and enough water to make twice as much liquid, in volume, as rice—about 4 cups [1 liter] of liquid—then add the peas, bay leaf and snails.

Turn up the heat, and cook over high heat until the liquid comes to a boil; then place the pan in a preheated 350° F. [180° C.] oven for 15 to 20 minutes. A few minutes before removing the paella from the oven, add the mussels. The paella should be served from the dish in which it was cooked.

CANDIDO LOPEZ
EL LIBRO DE ORO DE LA GASTRONOMIA

Paella Valenciana

The technique of cleaning squid is demonstrated on page 84. To prepare mussels, scrub the shells and scrape off the beards. Discard any mussels whose shells have opened.

The volatile oils in hot chilies may make your skin sting and your eyes burn; after handling chilies, avoid touching your face and wash your hands promptly.

To serve 8 to 10

2½ cups	unprocessed white rice	625 ml.
2 to 2½ lb.	chicken, cut into serving pieces	1 to 1¼ kg.
½ cup	oil	125 ml.
	salt	
½ lb.	lean pork, chopped	¼ kg.
½ lb.	fresh squid, cleaned and chopped	¼ kg.
¾ cup	chopped onion	175 ml.
1	red pepper, broiled, peeled, seeded and chopped	1
3	small tomatoes, peeled, seeded and chopped	3
½ lb.	live mussels	¼ kg.
½ lb.	chorizo or other spicy frying sausage	¼ kg.
½ lb.	fresh shrimp	¼ kg.
1 cup	green beans, cut into 2-inch [5-cm.] pieces	¼ liter
1¼ cups	shelled fresh peas, parboiled for 5 minutes in salted water and drained	300 ml.
4	garlic cloves	4
	powdered saffron	

In a skillet, begin browning the chicken in 2 tablespoons [30 ml.] of the oil; season it with salt. After about 10 minutes, when the chicken is half-cooked, add the pork and squid. Cover and cook slowly for 10 minutes. Add the onion and pepper; when the vegetables begin to brown, add the tomatoes. Simmer the mixture, uncovered, until the tomatoes have reduced to a pulp—about 10 minutes—then add the mussels and sausage. Cover and cook for about three min-

utes more, or until the mussels have opened. Remove the pan from the heat.

Heat ¼ cup [50 ml.] of the remaining oil in a paella pan or large skillet. Add the rice and fry it until it begins to color. Then add the chicken-and-sausage mixture. Stir and pour in 2½ cups [625 ml.] of boiling water. Add the shrimp, green beans and peas. Pound the garlic with a pinch of powdered saffron and the remaining oil. Add this paste to the paella, cover, and cook for a further 15 minutes. Set the paella aside to rest for five minutes, then serve it from the pan.

VICTORIA SERRA
TIA VICTORIA'S SPANISH KITCHEN

Ham Jambalaya

The technique of making jambalaya, using blue crabs and fresh oysters, is demonstrated on page 88.

To prepare chicken or turkey jambalaya, substitute chopped cooked chicken or turkey for the ham.

For seafood jambalaya, omit the oil and the ham. Sauté ½ pound [¼ kg.] of sliced *chorizo* or other hot pork sausage until the fat is rendered. Add the onions, green pepper and garlic. Substitute fish stock *(recipe, page 166)* or water for the chicken stock, and stir in 2 cups [½ liter] of shucked oysters or 1 pound [½ kg.] of crab meat, shrimp or rock lobster cut into chunks—or a combination of the shellfish.

To serve 4

1½ cups	unprocessed white rice	375 ml.
2 tbsp.	oil	30 ml.
2	large onions, chopped	2
1	green pepper, chopped	1
2	garlic cloves, finely chopped	2
1 cup	cubed cooked ham	¼ liter
1	bay leaf, crumbled	1
½ tsp.	dried thyme leaves	2 ml.
½ tsp.	salt	2 ml.
3 to 4 drops	Tabasco sauce	3 to 4 drops
3	tomatoes, peeled, seeded and chopped	3
4 cups	chicken stock *(recipe, page 166)*	1 liter

Heat the oil in a large, heavy casserole. Add the onions, green pepper and garlic, and cook, stirring now and then, until the onion bits are golden—about seven minutes. Add the ham and rice, and cook and stir until the rice is well coated with oil. Add the bay leaf, thyme, salt, Tabasco sauce, tomatoes and stock. Cover the casserole, and simmer the jambalaya until the rice is tender and the liquid absorbed— 20 to 25 minutes. Before serving, taste the jambalaya and add more salt and Tabasco sauce if needed.

JEANNE A. VOLTZ
THE FLAVOR OF THE SOUTH

Pink Pilaf

Domatesli Pilaf

To serve 6 to 8

3 cups	unprocessed white rice	¾ liter
8 tbsp.	butter	120 ml.
3 cups	chicken, lamb or beef stock *(recipes, page 166)*	¾ liter
3 cups	tomato juice	¾ liter
	salt and pepper	

Melt the butter in a heavy frying pan. Stir in the rice and heat it until the butter bubbles. Mix the stock and tomato juice, boil and pour over the rice. Add salt and pepper and mix well. Bake, covered, in a preheated 375° F. [190° C.] oven for 30 minutes. Mix again and bake for 20 minutes.

GEORGE MARDIKIAN
DINNER AT OMAR KHAYYAM'S

Tomato Pilaf

Pilafi meh Domata

To serve 6

2 cups	unprocessed long-grain white rice	½ liter
1 tbsp.	butter	15 ml.
1 tbsp.	oil	15 ml.
2	onions, chopped	2
1	garlic clove, finely chopped	1
½	green pepper, seeded, deribbed and cut into slivers	½
1 lb.	raw or cooked meat, such as chicken, beef or pork, cut into small pieces	½ kg.
2	tomatoes, peeled, seeded and chopped	2
3 cups	chicken stock *(recipe, page 166)*	¾ liter
½ cup	dry white wine	125 ml.
2 tsp.	salt	10 ml.
1 tsp.	sugar	5 ml.
	pepper	
1 lb.	raw or cooked mixed seafood, such as shelled shrimp, shucked clams, lobster meat and crab meat	½ kg.

Mix the butter and oil in a large skillet, and fry the onions, garlic and green pepper in it for five minutes over medium heat. If the meat is raw, add it to the skillet mixture for 10 minutes. Add the tomatoes, stock, wine, salt, sugar and a

little pepper. Bring to a boil. Add the rice, stir and cover the skillet. Reduce the heat and simmer for 20 minutes. Stir once or twice to prevent sticking. Add the raw seafood and, if using, the precooked meat. Cook the mixture 10 minutes more. If using cooked seafood, add it now and cook an additional three minutes. Serve hot.

THERESA KARAS YIANILOS
THE COMPLETE GREEK COOKBOOK

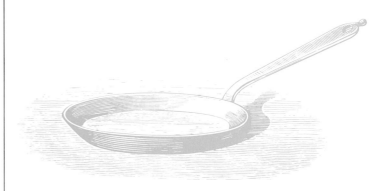

Yellow Rice, Nepalese-Style

To make the coconut milk specified in this recipe, first puncture the three eyes at one end of the coconut shell, drain out the liquid and split the shell in half. Pry the white meat from the shell with a small knife, pare off the papery brown skin, then grate the meat to make 3 to 4 cups [¾ to 1 liter]. Soak the meat for 30 minutes in an equal volume of boiling water; drain the liquid through a cheesecloth-lined colander set over a bowl, then squeeze the meat in the cheesecloth to extract all of the liquid. In a covered container, the milk will keep in the refrigerator for a day or two.

To serve 4

1 cup	unprocessed white rice, soaked for 30 minutes and drained	¼ liter
¼ cup	oil	50 ml.
8	scallions, sliced	8
2-inch	cinnamon stick	5-cm.
3	cardamom pods	3
4	whole cloves	4
	turmeric	
2 cups	coconut milk	½ liter
	salt	
3 or 4	sprigs coriander	3 or 4

Heat the oil in a skillet and fry the scallions until they are brown and crisp. Remove the scallions and drain them on paper towels. Add the spices, rice and a pinch of turmeric to the pan, and fry them for three minutes. Add 1 cup [¼ liter] of hot water, the coconut milk and a little salt. Cover the pan and cook the mixture over low heat for about 20 minutes, until the rice is tender. Add the coriander, and place the pan, uncovered, in a preheated 300° F. [150° C.] oven for five to 10 minutes to evaporate any excess moisture. Serve hot, garnished with the fried scallions.

E. MAHESWARI DEVI
HANDY RICE RECIPES

Spinach and Rice
Spanakorizo

A similar dish, *prassorizo*, is made with leeks and rice: Substitute 1½ pounds [¾ kg.] of sliced leeks for the spinach, and add the leeks at the same time as the liquid.

To serve 6 to 8

½ cup	unprocessed long-grain white rice	125 ml.
2 tbsp.	clarified butter or olive oil	30 ml.
1	small onion, chopped	1
½ cup	tomato sauce (recipe, page 167)	125 ml.
2 lb.	spinach, stems removed, washed and drained	1 kg.
¼ cup	chopped fresh parsley	50 ml.
2	sprigs mint, chopped	2
	salt and freshly ground pepper	
	freshly grated nutmeg	
1	lemon, cut into wedges	1
3 oz.	bacon, fried, drained and crumbled (optional)	90 g.
4	eggs, hard-boiled and sliced (optional)	4

Heat the butter or oil in an enameled pan, then add the onion and cook over low heat until soft and transparent. Add the rice and sauté for a few minutes, stirring constantly, then add the tomato sauce and about ¼ cup [50 ml.] of water.

Cover the pan and simmer until the rice is almost tender—about 10 minutes. Uncover the pan and stir in the spinach, parsley and mint; season with salt and pepper.

Partially cover the pan and continue cooking, stirring with a wooden spoon, until the spinach has wilted. Grate a little nutmeg over the top, and continue cooking until all of the liquid has been absorbed and the *spanakorizo* is tender, but not mushy.

Remove the pan from the heat and drape it with a dry towel, set beneath the lid, until ready to use. Transfer the *spanakorizo* to a warmed serving dish, and garnish with the lemon wedges, and, if you like, the crumbled bacon bits and sliced eggs. Serve warm.

VILMA LIACOURAS CHANTILES
THE FOOD OF GREECE

Lamb Pilaf with Double Stock
Yakhni Pilau

The technique of making this dish is demonstrated on pages 82-83. To prepare double stock, simmer ½ pound [¼ kg.] of lean lamb and such gelatinous cuts as a shank in about 4 cups [1 liter] of meat stock (recipe, page 166) for one and one half hours. Strain and degrease. The stock may be flavored with a few strips of lemon peel, parsley, coriander leaves, 1 tablespoon [15 ml.] of peeled and grated fresh ginger, a 2-inch [5-cm.] cinnamon stick, a fresh green chili or sweet green pepper, 1 teaspoon [5 ml.] of peppercorns and a few chopped chives.

To serve 4

1 cup	unprocessed long-grain white rice, soaked for 1 hour and drained	¼ liter
1¼ lb.	boned lamb, cubed	600 g.
6 tbsp.	strained fresh lime juice	90 ml.
8	anise or fennel seeds, ground	8
⅔ cup	heavy cream, lightly whipped	150 ml.
⅔ cup	yogurt	150 ml.
8 tbsp.	clarified butter	120 ml.
4	cloves, lightly crushed	4
2½ cups	double stock	625 ml.
7	cardamom pods, lightly crushed	7
2-inch	cinnamon stick	5-cm.
2 tbsp.	ground poppy seeds	30 ml.
2	bay leaves	2
	salt	
6 tbsp.	chopped fresh spinach or watercress	90 ml.

Prick the lamb cubes all over with a fork and rub them with the lime juice and anise or fennel. Blend the cream and yogurt and pour over the lamb. Set the lamb aside to marinate.

Heat 4 tablespoons [60 ml.] of the butter in a fireproof casserole and add the cardamom, cinnamon, poppy seeds and bay leaves. Add the rice and cook over medium heat, stirring until the rice becomes opaque—six to eight minutes.

In a saucepan, heat 2 tablespoons [30 ml.] of the butter. Add the cloves and cook for a few moments, then add the stock and stir. Cover the pan and cook over low heat for five minutes. Strain the marinade from the meat and add it to the rice, then add enough double stock to cover the rice by 1¼ inches [3 cm.]. Bring the liquid to a boil, season with salt and add the spinach or watercress.

Cover the casserole and bake in a preheated 350° F. [180° C.] oven for 12 minutes. Meanwhile, fry the lamb in the last of the butter and, when the meat is well browned, mix it into the rice. Cover the pilaf, reduce the oven heat to 325° F. [160° C.], and bake for 15 minutes, until the rice is tender. Leave the pilaf in a warm place for five minutes before serving.

DHARAMJIT SINGH
INDIAN COOKERY

Mussel Pilaf

Pilaf de Moules

To prepare mussels, scrub the shells and scrape off the beards. Discard any mussels whose shells have opened.

To make a hot pilaf, substitute 6 tablespoons [90 ml.] of heavy cream for the vinegar dressing; stir the cream in just before serving.

To serve 2 to 4

1¼ cups	unprocessed long-grain white rice	300 ml.
2 quarts	live mussels, scrubbed	2 liters
⅔ cup	dry white wine	150 ml.
1	garlic clove, crushed (optional)	1
2 tbsp.	butter	30 ml.
1	onion, thinly sliced	1
	salt and freshly ground pepper	
2 tbsp.	chopped fresh parsley	30 ml.
Vinegar dressing		
1 tbsp.	white wine vinegar	15 ml.
3 tbsp.	oil	45 ml.
	salt and freshly ground pepper	

Put the mussels in a large pan, add the white wine and garlic, if included; cover the pan and cook over high heat, stirring once, for five to seven minutes, or until the mussels have opened.

Remove the mussels from their shells. Strain the cooking liquid through cheesecloth, and add enough water to it to make 2½ cups [625 ml.].

Melt the butter in a heavy-based pan, and cook the onion slowly until soft but not brown. Add the rice and cook, stirring, until the grains are opaque and the butter is absorbed. Add the mussel liquid with some salt, if necessary, and pepper. Cover the pan and bring the liquid to a boil.

Transfer the pan to a preheated 350° F. [180° C.] oven and cook for 15 minutes.

Stir in the mussels, cover, and continue cooking for a further five minutes. After removing the pan from the oven, let the rice stand for five to 10 minutes. The pilaf can be cooked up to eight hours ahead and stored, covered, in the refrigerator.

To make the dressing, combine the vinegar, oil, and salt and pepper to taste. One to two hours before serving, gently stir the dressing into the rice and taste the salad for seasoning. Cover and leave in the refrigerator for the flavors to blend. Just before serving, stir in the parsley.

ANNE WILLAN
FRENCH COOKERY SCHOOL

Pilaf with Pine Nuts

To serve 6

2 cups	unprocessed long-grain white rice	½ liter
10 tbsp.	butter	150 ml.
	salt	
⅓ cup	dried currants	75 ml.
½ cup	pine nuts or blanched, peeled and slivered almonds	125 ml.
2 tbsp.	chopped fresh parsley	30 ml.
1 tbsp.	finely slivered orange peel	15 ml.

Melt 2 tablespoons [30 ml.] of the butter in a large saucepan, add the rice, and stir until the grains are coated. Pour in 4 cups [1 liter] of boiling salted water, cover, and simmer for 25 minutes, or until the rice is tender. Add the currants, fluff the rice with a fork and remove the pan from the heat.

In a small pan, melt the remaining butter, add the nuts, and heat them until they are lightly toasted and the butter starts to brown; pour the nuts and butter over the rice, and mix them gently. Transfer the rice to a serving dish and sprinkle with the parsley and orange peel.

JOSÉ WILSON (EDITOR)
HOUSE & GARDEN'S PARTY MENU COOKBOOK

Ground Meat and Savory Rice

Kheema Biryani

The volatile oils in hot chilies may make your skin sting and your eyes burn; after handling chilies, avoid touching your face and wash your hands promptly.

To serve 6 to 8

Saffron rice		
2 cups	unprocessed white rice	½ liter
5	peppercorns	5
⅓ tsp.	cumin seeds	1½ ml.
⅓ tsp.	coriander seeds	1½ ml.
1	thick slice fresh ginger, crushed	1
four 1-inch	cinnamon sticks	four 2½-cm.
8	whole cloves	8
8	cardamom pods, crushed	8
2	bay leaves	2
2½ cups	chicken or beef stock *(recipes, page 166)*	625 ml.
2 tsp.	salt	10 ml.
¼ tsp.	saffron threads	1 ml.

Spiced meat

4 tbsp.	*ghee*	60 ml.
2	medium-sized onions, finely chopped	2
3	garlic cloves, finely chopped	3
1½ tsp.	peeled and grated fresh ginger	7 ml.
2	bay leaves	2
	cumin seeds	
2 lb.	ground lamb or turkey	1 kg.
½ tsp.	turmeric	2 ml.
½ tsp.	freshly grated nutmeg	2 ml.
½ tsp.	ground mace	2 ml.
½ tsp.	ground cardamom seeds	2 ml.
½ tsp.	ground cinnamon	2 ml.
¾ tsp.	ground cumin	4 ml.
¾ tsp.	salt	4 ml.
1 cup	yogurt or sour cream	¼ liter
2 to 3 tbsp.	chopped fresh coriander leaves	30 to 45 ml.

Nut sauce

6 tbsp.	*ghee*	90 ml.
1	large onion, thinly sliced	1
2	fresh hot green chilies, stemmed, seeded and halved	2
2 oz.	almonds, blanched and peeled (about ½ cup [125 ml.])	60 g.
1 oz.	cashews (about ¼ cup [50 ml.])	30 g.
½ cup	raisins	125 ml.
1¼ cups	chicken or beef stock *(recipes, page 166)*	300 ml.
2½ tbsp.	rose-flower water	37 ml.

For the saffron rice, tie together in a 4-inch [10-cm.] square of cheesecloth, the peppercorns, cumin, coriander, ginger, cinnamon sticks, cloves, cardamom and bay leaves. In a large saucepan, bring the stock to a rapid boil. Add the salt, spice bag and saffron, and sprinkle in the rice. Boil over moderately high heat for two to three minutes. Reduce the heat to medium, cover the pan, and simmer for 10 minutes. Remove the pan from the heat, let the rice stand for five minutes, and fluff it with a fork. Set aside until ready to use.

To prepare the spiced meat, heat the *ghee* in a skillet. Add the onions, garlic, ginger, bay leaves and a pinch of cumin seeds; cook the mixture over medium heat for about two minutes. Stir in the ground meat, turmeric, nutmeg, mace, cardamom, cinnamon, ground cumin and salt. Cook the meat gently for eight to 10 minutes, or until most of the cooking juices have evaporated. Stir in the yogurt or sour

cream and the coriander leaves. Remove the spiced meat from the heat and set aside.

To assemble the dish, heat 3 tablespoons [45 ml.] of the *ghee* in a large casserole with a tight-fitting lid. Brown the sliced onion, then add the chilies, almonds, cashews, raisins and saffron rice. Stir the ingredients together for one or two minutes over medium heat.

Remove two thirds of the rice mixture, and spread the remaining rice evenly in the bottom of the casserole. Spoon one half of the spiced meat over the rice. Add another layer of rice, using one half of that remaining, cover it with the remaining spiced meat, and end with a layer of rice. Pour on the stock and the remaining 3 tablespoons of the *ghee;* sprinkle with rose water.

Cover the casserole and bake in a preheated 375° F. [190° C.] oven for 30 to 35 minutes, or until the rice has absorbed the liquid. Fluff the rice with a fork. Garnish, if you like, with additional chopped coriander leaves, seeded green chilies, tomato wedges and green pepper rings.

PRANATI SEN GUPTA
THE ART OF INDIAN CUISINE

Salt Cod Pudding
Kabiljo Pudding
To serve 4 to 6

2 cups	unprocessed white rice	½ liter
1 lb.	salt cod, soaked overnight and drained	½ kg.
2 cups	milk	½ liter
11 tbsp.	butter, melted	165 ml.
½ tsp.	white pepper	2 ml.
	salt	
1 to 2 tsp.	sugar	5 to 10 ml.
2	eggs, lightly beaten	2
about 5 tbsp.	fine dry bread crumbs	about 75 ml.

Place the fish in a large saucepan of water. Bring to a boil and immediately turn the heat as low as possible. Poach the fish for 10 to 12 minutes, or until flaky. Drain and, when cold again, bone and finely chop the fish.

Put the rice in 2 cups [½ liter] of cold water and, when boiling, add the milk. Cover the rice and simmer until almost soft—about 20 minutes; add 3 tablespoons [45 ml.] of the butter and let the rice cool. Season with pepper, salt and sugar; add the fish and beaten eggs. Pour the mixture into a lightly buttered pie dish coated with about 3 tablespoons of bread crumbs, sprinkle with the remaining bread crumbs, and bake in a preheated 375° F. [190° C.] oven for about an hour, or until a wooden pick stuck in the center emerges clean. Serve with additional melted butter.

INGA NORBERG
GOOD FOOD FROM SWEDEN

Old-fashioned Rice Pudding

The technique of making this dish is demonstrated on page 68. This pudding can be flavored with ¼ cup [50 ml.] of mixed dark and golden raisins, ¼ cup of chopped nuts and one 3-inch [7½-cm.] cinnamon stick. To give the pudding a sweet crust, blend ½ cup [125 ml.] of light brown sugar into 2 tablespoons [30 ml.] of slightly softened butter until the mixture crumbles easily. Scatter the mixture over the cooked pudding and place under the broiler for approximately five minutes, until the sugar has partly melted.

To serve 4 to 6

½ cup	unprocessed white rice	125 ml.
4 cups	milk	1 liter
½ cup	sugar	125 ml.
½ tsp.	salt	2 ml.

Mix all of the ingredients, turn them out into a buttered baking dish, and bake, covered, in a preheated 325° F. [160° C.] oven until the rice has softened—about one and one half hours. Uncover the dish to brown the top surface slightly.

THE SETTLEMENT COOKBOOK

Kedgeree

The techniques of cooking rice are explained on pages 22-23.

Kedgeree was a famous breakfast or supper dish in England in the 18th and 19th Centuries. The name comes from the Hindi, but the parent Indian dish is quite different.

To serve 4

2 cups	cooked long-grain white rice	½ liter
6 oz.	smoked haddock, poached, bones and skin removed, finely flaked	180 g.
2	eggs, hard-boiled and finely chopped	2
	salt and freshly ground pepper	
about 5 tbsp.	butter	about 75 ml.
1 tbsp.	finely chopped fresh parsley or mixed parsley and chives	15 ml.

Lightly mix the rice, fish and eggs, and season them with a little salt and plenty of freshly ground pepper. Melt 2 to 2½ tablespoons [30 to 37 ml.] of the butter in a saucepan. Stir in the rice mixture and toss it until it is heated through; alternatively, put the mixture in a covered dish with plenty of butter in a preheated 325° F. [160° C.] oven and bake for 20 minutes. To serve, pile the kedgeree in a flat dish, place the remaining butter, cut into pieces, on the pile, and sprinkle with the parsley or parsley and chives. Serve very hot with toast and butter.

ELISABETH AYRTON
THE COOKERY OF ENGLAND

Eggplant, Zucchini and Peppers Baked with Rice

Ratatouille en Estouffade

This is delicious made the day before and then reheated. If any is left over, it is also very good cold. It can be served from the casserole or from a serving platter. If you wish to gratiné it, turn it into a large oval gratin dish, sprinkle it with ½ cup [125 ml.] of mixed grated Gruyère and Parmesan cheeses, dribble olive oil over it, and bake in a preheated 400° F. [200° C.] oven for 15 to 20 minutes, until it is hot through. Then broil it until the cheese is bubbling and brown.

To serve 8 to 10

½ cup	unprocessed long-grain white rice	125 ml.
½ cup	olive oil	125 ml.
5	garlic cloves, finely chopped	5
4	medium-sized onions, sliced	4
1	medium-sized eggplant, peeled and cut into ½-inch [1-cm.] cubes	1
3 or 4	medium-sized zucchini, sliced into rounds ⅛ inch [3 mm.] thick	3 or 4
5 or 6	medium-sized red or green peppers, seeded, deribbed and cut into strips	5 or 6
9	medium-sized tomatoes, peeled, seeded and chopped, or two (35-oz. [1-kg.]) cans Italian-style plum tomatoes, drained, seeded and chopped	9
3 tbsp.	mixed chopped fresh parsley and basil leaves	45 ml.
	salt and pepper	
½ cup	chicken stock (recipe, page 166)	125 ml.

Stir the rice into 2 quarts [2 liters] of boiling salted water and cook it for eight minutes. Drain the rice and refresh it under cold running water. Set it aside to drain thoroughly.

Heat the oil in a 6- to 8-quart [6- to 8-liter] heatproof casserole with a lid. Add the garlic and onions, and gently sauté them over medium-low heat for six to eight minutes, stirring occasionally until the onions are beginning to be tender. Stir in the eggplant. After two minutes, stir in the zucchini and cook for 20 minutes, stirring frequently.

Remove two thirds of the vegetables from the casserole. Spread the remaining vegetables in a layer, and add alternating layers of the red or green peppers, tomatoes, rice and the cooked vegetables until all of the ingredients are used up. Sprinkle with herbs and season each layer as you go.

Pour on the chicken stock, cover the casserole and bake it in a preheated 350° F. [180° C.] oven for one hour; the vegetables should be very tender, but not a purée.

SIMONE BECK
NEW MENUS FROM SIMCA'S CUISINE

Rice Timbale from Lombardy
Timballo di Riso

The technique of preparing rice timbale is demonstrated on pages 62-63.

To serve 6 to 8

2⅔ cups	unprocessed long-grain white rice	650 ml.
8 tbsp.	butter, cut into pieces, at room temperature	120 ml.
⅓ cup	freshly grated Parmesan cheese	75 ml.
4	egg yolks	4
	fine dry bread crumbs	
	tomato sauce (recipe, page 167)	

Meat filling

2 tbsp.	butter	30 ml.
¼ cup	finely chopped onion	50 ml.
1	garlic clove, finely chopped	1
½ lb.	boneless lean veal, ground twice	¼ kg.
½ lb.	boneless lean pork, ground twice	¼ kg.
½ lb.	chicken livers, chopped	¼ kg.
1	small red pepper, seeded, deribbed and chopped	1
1 cup	shelled fresh peas, parboiled in water for 3 minutes	¼ liter
¼ cup	puréed tomato (recipe, page 166), flavored with a few fresh basil leaves	50 ml.
about 2 tbsp.	dry white wine	about 30 ml.
1 tsp.	salt	5 ml.
½ tsp.	sugar	2 ml.
½ tsp.	dried oregano leaves	2 ml.
¼ tsp.	freshly ground pepper	1 ml.
	Tabasco sauce	

Cook the rice in plenty of boiling salted water for about 15 minutes, or until it is almost but not quite tender. Drain the rice well and place it in a bowl. Mix in the butter, Parmesan and egg yolks.

Generously butter a 3-quart [3-liter] mold or casserole and thoroughly coat the bottom and sides of it with bread crumbs; there must not be any uncovered spots or the timbale will not unmold. Spoon two thirds of the rice mixture into the casserole. Press the rice against the bottom and sides, leaving a well in the middle.

To make the filling, heat the butter in a deep skillet. Add the onion and garlic, and cook, stirring constantly, until the onion is soft and golden. Add the veal and pork; stir well to mix the meats and blend them. Stir in the chicken livers.

Cook the meats over low heat, stirring frequently for about 15 minutes, or until they are cooked through. Add the other filling ingredients and mix well. Then, stirring often, cook the filling over low heat for about 15 minutes; if necessary, add a little more wine, a tablespoon [15 ml.] at a time, to prevent sticking. The filling must be very thick.

Spoon the filling into the well in the rice, pressing it into place and smoothing the surface. Spoon the remaining rice over the filling and smooth the top, taking care that the meat is well covered. Bake without a cover in a preheated 350° F. [180° C.] oven for about one hour, or until the rice is thoroughly set. Unmold carefully onto a heated platter, first running a knife blade around the edge of the mold to loosen the rice if necessary. Cut the timbale into wedges and serve immediately with tomato sauce on the side.

NIKA HAZELTON
THE REGIONAL ITALIAN KITCHEN

Rice with Pork and Vegetables
Arroz Murciano

To serve 6 to 8

2 cups	unprocessed white rice	½ liter
1 lb.	lean pork, cut into pieces	½ kg.
¾ cup	oil	175 ml.
2	garlic cloves	2
6	red peppers, seeded, deribbed and sliced	6
3	medium-sized tomatoes, peeled, seeded and chopped	3
3 tbsp.	chopped fresh parsley	45 ml.
½ tsp.	powdered saffron	2 ml.
	salt and freshly ground pepper	

In a large casserole, sauté the pork in the oil until browned, and set the pork aside. Over low heat, fry the garlic in the same oil until lightly colored, then remove it and put to one side. Put the peppers and tomatoes in the casserole and cook them for 10 minutes or so.

Pound the garlic to a paste with the parsley and saffron, and add this to the casserole with the peppers and tomatoes. Add the pork and a few tablespoons of water. Cook for a few minutes, or until the liquid has evaporated. Add the rice, stir it well and then pour in 5 cups [1¼ liters] of boiling water. Season the mixture with salt and pepper, and bring the water back to a boil. Cover the casserole, place it in a preheated 350° F. [180° C.] oven, and let the mixture cook for an hour.

MANUAL DE COCINA

Rice with Spinach and Ham

Reis mit Spinat und Schinken

The techniques of cooking rice are explained on pages 22-23.

To serve 4

3 cups	cooked white rice	¾ liter
1½ lb.	spinach, cooked and squeezed dry	¾ kg.
2 tbsp.	butter, softened	30 ml.
½ lb.	smoked ham, diced	¼ kg.
½ cup	shredded Gruyère cheese	125 ml.
about 2 cups	tomato sauce *(recipe, page 167)*	about ½ liter

Mix the spinach while it is still warm with the softened butter. Fill a buttered 9-inch [23-cm.] casserole or soufflé dish with alternate layers of rice, ham and spinach, ending with a layer of rice.

Sprinkle the top with half of the grated cheese and cover the cheese with buttered parchment paper. Place the pan in a water bath and cook in a preheated 350° F. [180° C.] oven for about 30 minutes.

Turn the casserole out onto a warmed serving dish, remove the paper, pour on the tomato sauce and sprinkle with the rest of the grated cheese.

LILO AUREDEN
WAS MÄNNERN SO GUT SCHMECKT

Squid Baked with Rice

Kalamarakia Pilafi

The techniques of cleaning and preparing squid are demonstrated on page 84.

To serve 4

1 cup	unprocessed long-grain white rice	¼ liter
1 lb.	medium-sized fresh squid, cleaned, body pouches sliced into rings ½ to 1 inch [1 to 2½ cm.] thick	½ kg.
¼ cup	olive oil	50 ml.
3	garlic cloves, sliced	3
¼ cup	dry white wine	50 ml.
2	tomatoes, peeled, seeded and sliced	2
3 tbsp.	butter	45 ml.
about 5 tbsp.	chopped fresh parsley	about 75 ml.
1 tbsp.	chopped fresh rosemary leaves or 1 tsp. [5 ml.] dried rosemary	15 ml.
	salt and freshly ground pepper	

Heat the oil in a skillet, add the squid and the garlic, and sauté them for five minutes. Stir in the wine and the toma-toes, cover the pan and simmer the mixture until the squid is almost tender—approximately 30 minutes. Transfer the mixture to a baking dish.

Meanwhile, in a separate pan, heat the butter and sauté the rice—stirring constantly to avoid browning it—until it is milky and opaque. Add the rice to the squid, and sprinkle with 4 tablespoons [60 ml.] of the parsley, the rosemary, and salt and pepper to taste. Add 1¼ cups [300 ml.] of hot water. Cover the baking dish and bake in a preheated 350° F. [180° C.] oven for 30 to 40 minutes, or until the rice is tender. Sprinkle with the remaining parsley and serve hot.

VILMA LIACOURAS CHANTILES
THE FOOD OF GREECE

Savory Rice

The techniques for cooking rice are explained on pages 22-23. The volatile oils in hot chilies may make your skin sting and your eyes burn; after handling chilies, avoid touching your face and wash your hands promptly.

To serve 2 to 4

2½ cups	cooked rice	625 ml.
2	leeks, chopped	2
3½ tbsp.	*ghee*	50 ml.
½	fresh hot green chili, stemmed, seeded and chopped	½
2 tsp.	chopped red onion	10 ml.
4	eggs, beaten	4
	salt and pepper	

Fry the leeks in the *ghee* for about five minutes, then add the chili and onion and fry them all together until they are soft but not brown. Season the eggs with salt and pepper, turn them into the frying pan, and fry everything together, scrambling the eggs into rather coarse pieces by mixing round and round with a wooden spoon.

Now add the cooked rice, a big spoonful at a time, mixing well after each addition. Pile up the eggs and rice on a hot dish and serve, or form the rice into a ring: Pack the rice firmly into a lightly buttered 2½- to 3-cup [675- to 750-ml.] ring mold, pressing down with the back of a spoon until it is just above the level of the top of the mold. Smooth off the top with the blade of a knife; then put the mold into a preheated 300° F. [150° C.] oven for about five minutes; turn the ring out onto a serving platter and serve hot.

HILDA DEUTROM (EDITOR)
CEYLON DAILY NEWS COOKERY BOOK

Yogurt, Lamb and Rice

Tah Chin

To serve 6

3 cups	short- or round-grain rice, preferably basmati, rinsed and drained	¾ liter
2 lb.	boned leg of lamb, cut into 1¼-inch [3-cm.] cubes	1 kg.
2½ cups	yogurt	625 ml.
2 tbsp.	salt	30 ml.
	freshly ground pepper	
1 tsp.	turmeric, or ½ tsp. [2 ml.] saffron threads, pulverized	5 ml.
2	egg yolks	2
4 tbsp.	ghee or butter, melted	60 ml.

In a bowl, combine the lamb with the yogurt, 2 teaspoons [10 ml.] of the salt, a good grinding of pepper and the turmeric or saffron. Cover the bowl and let the mixture marinate in the refrigerator for six hours or overnight.

Cook the rice, uncovered, for five minutes in a large pot of boiling salted water. Drain the cooked rice in a colander or in a large sieve.

Beat the egg yolks in a bowl; stir in ½ cup [125 ml.] of the yogurt marinade from the lamb and 1½ cups [375 ml.] of the cooked rice. Pour the melted *ghee* or butter and 1 tablespoon [15 ml.] of hot water in a 2½-quart [2½-liter] casserole and swirl the vessel to coat the sides. Spread the egg, yogurt and rice mixture evenly over the bottom of the casserole.

Arrange half of the lamb cubes over this with some more of the yogurt marinade. Add a second layer of rice, then the remaining lamb, reserving ½ cup of the yogurt marinade. Top with the remaining rice and spread the last of the yogurt marinade on top. Cover the casserole and cook in a preheated 325° F. [160° C.] oven for one and three quarters hours.

Spoon the rice-and-lamb mixture into the center of a serving dish. Lift off the crusty layer from the bottom of the dish and break it into large pieces. Arrange them around the edge of the dish and serve.

TESS MALLOS
THE COMPLETE MIDDLE EAST COOKBOOK

Malaysian Fried Rice

Nasi Goreng

The techniques of frying rice are explained on pages 22-23. The volatile oils in hot chilies may make your skin sting and your eyes burn; after handling chilies, avoid touching your face and wash your hands promptly.

This dish, one of the most common in Malaysia, always includes shrimp, which gives it a strong marine flavor. In other respects it is subject to small variations. The Chinese make it with pork, decorate it elaborately with slivers of omelet, chopped scallion, red chilies and parsley, and serve it with a thick chili sauce. The Malay people use beef as the meat, adding onions and sometimes green peas. They present the dish in a neat mold surrounded by shredded lettuce, decorate it with chilies cut into flower shapes and a single fried egg on top, and serve it with soy sauce containing thin slices of chili. The following recipe provides what many Malaysians would regard as generous quantities of shrimp and meat. Less may be used. There does exist in coastal villages a *nasi goreng* that contains fish in place of meat. I recommend trying it by substituting 5 ounces [150 g.] of chopped fish for the meat.

To serve 4

1 cup	cooked white rice	¼ liter
¼ lb.	pork or beef, trimmed of fat and finely chopped	125 g.
¼ lb.	fresh shrimp, shelled and deveined	125 g.
1	garlic clove, finely chopped	1
5	small onions, finely chopped (optional)	5
1 tbsp.	raisins	15 ml.
1 tbsp.	soy sauce	15 ml.

Garnishes

4	small red onions, sliced or chopped, and fried	4
4	lettuce leaves, shredded	4
2	fresh hot red chilies, stemmed, seeded and cut into slivers or flower shapes	2
3	scallions, finely chopped	3
1	one-egg flat omelet, cooked like a pancake, rolled up and cut into strips, or 1 fried egg	1
1 tbsp.	chopped fresh parsley	15 ml.
	croutons	

Fry the meat with the garlic, and onions if using, until brown; add the shrimp and stir for a couple of minutes until they turn pink. Stir in the cooked rice, raisins and soy sauce, and continue to fry gently until the mixture is thoroughly heated and evenly browned.

Garnish the dish according to taste with the red onions, lettuce, chilies, scallions, egg, parsley and croutons. Serve, if you wish, with chili, soy or Tabasco sauce.

ALAN DAVIDSON
SEAFOOD OF SOUTH-EAST ASIA

Fried Rice

The technique of making this dish is shown on pages 42-43.

To serve 4

4 cups	cooked long-grain white rice, at least 1 day old	1 liter
2 oz.	beef tenderloin	60 g.
¼	garlic clove, finely chopped	¼
	salt and pepper	
½ tsp.	cornstarch	2 ml.
½ tsp.	sugar	2 ml.
2 tbsp.	soy sauce	30 ml.
5 tbsp.	oil	75 ml.
2	slices peeled fresh ginger, cut into julienne	2
1	green pepper, seeded, deribbed and cut into julienne	1
2 oz.	bacon, diced	60 g.
6	scallions, thinly sliced	6
6 oz.	cooked ham, diced	175 g.
2 to 4 tbsp.	chopped fresh parsley or coriander leaves	30 to 60 ml.
6	eggs, lightly beaten	6

Partially freeze the beef to make it easier to carve. Slice it paper-thin. Toss the slices with the garlic, a little salt and pepper, the cornstarch, sugar, 1 tablespoon [15 ml.] of the soy sauce and 2 tablespoons [30 ml.] of the oil. Set the mixture aside to marinate.

Heat the remaining oil in a large skillet. Add the ginger, green pepper and a little salt and pepper, and fry for one minute. Loosen the rice with your hands to separate the grains, and sprinkle them into the pan. When the grains are hot and separated, push them to the outside edges of the pan. Toss the bacon, scallions and ham into the center and fry them for a moment; then thoroughly mix them into the rice. Again form a well in the center of the rice, drop in the beef and its marinade, and fry for 30 seconds; then add about 2 tablespoons [30 ml.] of the parsley or coriander. Fry for another 30 seconds, then mix thoroughly into the rice.

Form yet another well in the center of the rice, drop in the eggs, and fry them until they reach the consistency of creamy scrambled eggs; then mix the eggs into the rice.

Sprinkle the mixture with the remaining soy sauce. Transfer it to a warm serving dish, and serve garnished with the remaining parsley or coriander if desired.

DOREEN YEN HUNG FENG
THE JOY OF CHINESE COOKING

Fried Balls of Rice

To serve 4

½ cup	unprocessed white rice	125 ml.
1	tomato, peeled, seeded and chopped	1
1	small red onion, finely chopped	1
2	eggs	2
	toasted bread crumbs	
	oil for deep frying	

Put the rice in a pot with ¾ cup [175 ml.] of cold water, cover the pot, and bring the liquid to a boil. Reduce the heat to low, and simmer until all of the water has been absorbed—approximately 20 minutes.

Heat a little oil in a skillet, lightly fry the tomato and onion together, and then put them into the pot with the rice. Lightly beat one of the eggs, and add just enough of it to the rice to bind the mixture together, stirring it with a wooden spoon over very low heat. Turn the rice mixture onto a plate, cover it with a towel, and let it cool.

Divide the cooled mixture into equal-sized pieces, and roll them into balls about the size of lichees or walnuts. Lightly beat the remaining eggs, coat the balls of rice with them and with the bread crumbs and, in a deep or shallow pan of oil heated to 375° F. [190° C.], fry the balls until they are golden brown. Drain the balls on paper towels and serve, if you like, with a spicy sauce.

LILIAN LANE
MALAYAN COOKERY RECIPES

New Orleans Rice Cakes

Calas

To serve 4

½ cup	unprocessed white rice	125 ml.
¼ oz.	package active dry yeast	7½ g.
3	eggs, beaten	3
¼ cup	flour	50 ml.
½ cup	sugar	125 ml.
½ tsp.	salt	2 ml.
	freshly grated nutmeg	
	oil for deep frying	
	confectioners' sugar	

Cook the rice in 3 cups [¾ liter] of boiling water until it is very tender—about 30 minutes. Drain the rice and set it aside to cool.

Dissolve the yeast in 2 tablespoons [30 ml.] of warm water. Mix the yeast with the cold cooked rice and let the rice stand in a warm spot overnight. The following day, beat in the eggs, flour, sugar, salt and a pinch of nutmeg, adding more flour, if necessary, to make a thick batter.

Heat the fat to 370° F. [190° C.], or until a 1-inch [2½-cm.] cube of bread browns in 60 seconds. Drop the batter from a tablespoon into the hot fat, and fry the balls until golden brown; drain them on paper towels, sprinkle with confectioners' sugar and serve hot.

THE EDITORS OF AMERICAN HERITAGE
THE AMERICAN HERITAGE COOKBOOK

Rice Patties
Kotlety z Ryżu

To serve 6

1 cup	unprocessed white rice	¼ liter
2 cups	milk	½ liter
6 tbsp.	butter	90 ml.
½ oz.	dried mushrooms, soaked in ¼ cup [50 ml.] warm water for 30 minutes and drained	15 g.
2	eggs, lightly beaten	2
	salt and pepper	
1 tbsp.	finely cut fresh dill, or mixed fresh dill and parsley	15 ml.
about ½ cup	dry bread crumbs	about 125 ml.

Mushroom sauce

2 tsp.	butter	10 ml.
1 tbsp.	flour	15 ml.
2 to 3 tbsp.	sour cream	30 to 45 ml.

In a saucepan, heat the milk and 1 tablespoon [15 ml.] of the butter. Add the rice, cover the pan and simmer over low heat for about 10 minutes; place the saucepan in a preheated 350° F. [180° C.] oven for 30 minutes to finish steaming, then set it aside to cool.

Simmer the mushrooms in their soaking liquid for about 15 minutes. Drain them, reserving the liquid for the mushroom sauce; chop the mushrooms.

While the rice is cooling, cream 5 tablespoons [75 ml.] of the remaining butter and the eggs. Add the mushrooms and dill, then combine the mixture with the cooled rice. Season the mixture with salt and pepper, and shape it into patties. Roll the patties in bread crumbs and fry them in additional butter until they are golden brown; drain the patties on paper towels and keep them warm.

To prepare the mushroom sauce, combine the butter and flour, and cook them, stirring, over moderate heat to brown the flour. Add the sour cream and reserved mushroom liquid. Whisk the mixture until it thickens and is smooth. Serve this on the rice patties.

MARJA OCHOROWICZ-MONATOWA
POLISH COOKING

Seafood Sizzling Rice
San Hsien Kuo Pa

To vary the seafood stew, replace the lobster or crab meat with scallops and use such vegetables as carrots, broccoli, bamboo shoots, fresh peas and bok choy.

To serve 4

1	rice crust (recipe, page 165), broken into 2-inch [5-cm.] pieces	1
½ lb.	small fresh shrimp, shelled, deveined and halved lengthwise	¼ kg.
½	egg white, lightly beaten	½
1 tsp.	cornstarch	5 ml.
½ tsp.	salt	2 ml.
3 cups	peanut or corn oil	¾ liter
½ cup	sliced fresh mushrooms	125 ml.
¼ cup	sliced water chestnuts	50 ml.
½ cup	snow peas	125 ml.
1 cup	cooked lobster or king crab meat, cut into 1-inch [2½-cm.] pieces	¼ liter

Soy chicken sauce

1 cup	chicken stock (recipe, page 166)	¼ liter
½ tbsp.	sugar	7 ml.
½ tsp.	salt	2 ml.
⅛ tsp.	pepper	½ ml.
1 tbsp.	cornstarch	15 ml.
1 tbsp.	soy sauce	15 ml.
1 tbsp.	distilled white vinegar	15 ml.

Mix the sauce ingredients in a dish. Combine the shrimp, egg white, cornstarch and salt and, using your hand, mix them well. Refrigerate the shrimp for 30 minutes or more.

Heat the oil in a wok or deep fryer until very hot—about 400° F. [200° C.]. Fry the rice crusts, two pieces at a time, until light brown and crispy—about 10 seconds. Drain the crusts, set them on a heatproof platter, and place in a preheated 475° F. [250° C.] oven for seven to eight minutes.

Let the oil cool to about 280° F. [145° C.]. Add the shrimp, and quickly stir them until they separate and change color. Pour the shrimp and oil into a strainer set over a pot.

Heat 2 tablespoons [30 ml.] of the oil in the wok or a skillet over medium heat. Stir fry the mushrooms, water chestnuts and snow peas for two minutes. Pour in the sauce, stirring until it forms a clear, light glaze. Increase the heat, add the lobster or crab meat and the shrimp, mix and heat through. Pour into a serving bowl. At the table, pour the seafood over the rice crusts; it will sizzle. Serve them at once.

FLORENCE LIN
FLORENCE LIN'S CHINESE REGIONAL COOKBOOK

Meat and Seafood Sizzling Rice

Hsia Jen Kuo Pa

Various substitutions are possible in this dish. Scallops, lobster meat or crab meat may be used instead of shrimp. Four ounces [125 g.] of veal, beef or chicken may replace the pork. Broccoli and asparagus may be used instead of the mushrooms and zucchini. Presoaked Chinese dried black mushrooms, water chestnuts, bamboo shoots or snow peas may be stir fried along with the other vegetables.

To serve 6

1	rice crust (recipe, page 166), broken into 2-by-2-inch [5-by-5-cm.] pieces	1
1 lb.	fresh small shrimp, shelled and deveined	½ kg.
2	boneless pork chops, cut ½ inch [1 cm.] thick and thinly sliced	2
1 cup	sliced fresh mushrooms	¼ liter
1½ cups	thinly sliced zucchini	375 ml.
6 tbsp.	peanut or corn oil	90 ml.
Shrimp marinade		
1½ tbsp.	egg white	22 ml.
2 tsp.	cornstarch	10 ml.
½ tsp.	salt	2 ml.
Pork marinade		
1 tsp.	cornstarch	5 ml.
1 tbsp.	water	15 ml.
1 tbsp.	soy sauce	15 ml.
1 tbsp.	dry sherry	15 ml.
Chicken sauce		
1 cup	chicken stock (recipe, page 166)	¼ liter
½ tsp.	salt	2 ml.
⅛ tsp.	white pepper	½ ml.
1 tbsp.	cornstarch	15 ml.
1 tbsp.	soy sauce	15 ml.

Combine the shrimp with the shrimp-marinade ingredients; mix well, cover, and refrigerate for at least 30 minutes or as long as 24 hours.

Combine the pork slices with the pork-marinade ingredients; mix well.

Heat a wok or skillet over medium heat until it is very hot. Add 4 tablespoons [60 ml.] of the oil, and stir fry the shrimp until they become firm and most of their flesh becomes pink. Remove them with a slotted spoon and set aside on a plate. Using the same wok or skillet and the left-over oil, stir fry the pork; if it is too dry, add some more oil. Remove the pork and set it aside with the cooked shrimp.

In very hot oil (about 400° F. [200° C.]), fry the rice crust two pieces at a time for about five seconds on each side. The rice pieces will puff up, will double in size, and will be light brown and crispy. Drain the puffed rice, set the pieces on a heat-proof serving platter, and place it in a preheated 450° F. [230° C.] oven for seven or eight minutes.

Meanwhile, heat a clean wok or skillet over medium heat with the remaining 2 tablespoons [30 ml.] of oil. Stir fry the mushrooms and zucchini for two minutes. Combine the sauce ingredients, stirring to make sure the cornstarch is well mixed. Slowly pour the sauce into the pan, stirring until it forms a light, clear glaze. Turn the heat to high, then add the cooked shrimp and pork, stirring quickly until they are just heated through. Transfer the shrimp-and-pork mixture to a hot serving dish.

At the dinner table, pour the cooked shrimp-and-pork mixture over the puffed rice. It will make a sizzling noise. Serve the dish immediately.

FLORENCE LIN
FLORENCE LIN'S CHINESE ONE-DISH MEALS

Sizzling Rice Soup

To serve 4

½	rice crust (recipe, page 165), broken into 3-inch [7½-cm.] pieces	½
5 cups	chicken stock (recipe, page 166) or water	1¼ liters
½ lb.	fresh shrimp, shelled and deveined	¼ kg.
¼ lb.	mushrooms, halved	125 g.
½ tsp.	salt	2 ml.
1 tbsp.	soy sauce	15 ml.
1 tbsp.	pale dry sherry	15 ml.
	pepper	
	oil for deep frying	

Bring the stock or water to a boil in a saucepan. Add the shrimp, mushrooms, salt, soy sauce, sherry and a pinch of pepper, and stir well. Cover the pan and simmer until the ingredients are heated through; pour the soup into a warmed tureen and keep it hot while frying the rice-crust pieces.

Heat the oil to 350° F. [180° C.]. Deep fry the crust pieces for about 30 seconds, or until crisp but not brown. Drain, and arrange them on a warmed plate.

To serve, place the crust pieces in individual bowls and, at the table, pour the stock mixture over them.

MARY MA STAVONHAGEN
THE COMPLETE ENCYCLOPEDIA OF CHINESE COOKING

Rice Timbale Stuffed with Sausages

Timballo di Riso

The techniques of assembling and cooking this timbale are demonstrated on pages 62-63. The dish may be accompanied by a sauce made by combining 1 cup [¼ liter] of meat stock (recipe, page 166) and ½ cup [125 ml.] of the soaking liquid from the dried mushrooms and boiling the mixture until it is reduced to one third of its original volume. Then enrich the sauce by adding 1½ cups [375 ml.] of heavy cream, and reduce it over high heat to one half of its original volume.

To serve 6

3 cups	short- or round-grain white rice	¾ liter
3	eggs	3
3 tbsp.	freshly grated Parmesan cheese	45 ml.
	salt and freshly ground pepper	
	freshly grated nutmeg	
½ cup	fresh bread crumbs	125 ml.

Sausage filling

2 tbsp.	olive oil	30 ml.
1 tsp.	butter	5 ml.
1	medium-sized red onion, coarsely chopped	1
6	sweet Italian sausages, peeled, and broken into small pieces	6
6	medium-sized tomatoes, peeled and seeded	6
1 cup	meat or chicken stock (recipes, page 166)	¼ liter
⅔ cup	dried Italian mushrooms, soaked in ¾ cup [175 ml.] warm water for 30 minutes and drained, stems removed and caps cut into pieces, the soaking liquid reserved	150 ml.

Put the rice in a saucepan with a large quantity of cold water and a pinch of salt. Set the saucepan on medium heat and stir the rice with a wooden spoon until the water reaches the boiling point. Reduce the heat and simmer the rice until it is half-cooked—about 10 minutes. Remove the saucepan from the heat and drain the rice. Run cold water over it to cool it completely, then drain it again and put it in a bowl. Add the eggs, Parmesan, salt, pepper and a pinch of grated nutmeg. Stir the mixture well until it is completely combined. Set it aside while you prepare the filling.

Heat the olive oil and the butter in a saucepan. Add the chopped onion and sauté gently for 10 to 12 minutes, or until it is golden brown. Add the sausage pieces and sauté for 10 to 15 minutes, then add the tomatoes and simmer very slowly for 15 minutes more. Add the stock, taste for salt and pepper, and let the mixture cook slowly until the stock has complete-ly evaporated and the mixture is thick and homogeneous (about 30 minutes).

Add the cut-up mushrooms and 3 to 4 tablespoons [45 to 60 ml.] of the soaking water. Continue cooking until the water has evaporated, then remove the saucepan from the heat and transfer the filling to another bowl to cool for about 30 minutes.

When cool, transfer 2 to 3 tablespoons [30 to 45 ml.] of the liquid from the filling mixture to the bowl with the rice.

To assemble the timbale, butter the bottom and sides of a 10-inch [25-cm.] round casserole and sprinkle it with some of the bread crumbs. Stir the rice very well once more and cover the bottom and sides of the casserole with three quarters of it. Pour the filling mixture into the center, then make a layer on top with the rest of the rice. Sprinkle the remaining bread crumbs over, then place the casserole in a preheated 400° F. [200° C.] oven for 25 to 30 minutes, or until the timbale is firmly set.

Remove the timbale from the oven, let it cool for 15 minutes and then unmold it onto a serving dish. Slice it like a cake and serve hot.

GIULIANO BUGIALLI
THE FINE ART OF ITALIAN COOKING

Rice Balls

Supplì

The technique of making this dish is shown on pages 56-57.

To serve 4 to 6

3 cups	cooked short- or round-grain white rice, or risotto	¾ liter
2	eggs, lightly beaten	2
¼ lb.	cooked ham or mortadella, thinly sliced and cut into small squares	125 g.
¼ lb.	mozzarella cheese, thinly sliced and cut into small squares	125 g.
	fine dry bread crumbs	
	oil or fat for deep frying	

Stir the eggs into the rice to bind it. Take about 1 tablespoon [15 ml.] of the rice and put it flat on the palm of your hand; on the rice, lay a little square of ham or mortadella and one of cheese. Place another tablespoon of rice on top of the meat and cheese, and form the mixture into a ball the size of a large Brussels sprout so that the meat and the cheese are completely enclosed. Roll each *supplì* very carefully in bread crumbs, then deep fry them in hot oil or fat heated to 375° F. [190° C.], turning them over and around so that the whole of the outside is nicely browned. Drain them on paper towels. The cheese should be just melted, stretching into threads (to which the dish owes its nickname of *supplì al telefono*).

ELIZABETH DAVID
ITALIAN FOOD

Salmon Pudding with Rice

Laxpudding med Risgryn

The salted salmon specified in this recipe is not generally available in the United States. Substitute fresh salmon or make your own salted salmon by rubbing coarse salt and a little sugar into all surfaces of a ¾-pound [350-g.] salmon fillet. Place the salmon on a plate, cover tightly with plastic wrap or foil, and weight with a board topped with several pounds of weights or several large cans of food. Refrigerate it for three days. Before using, drain the salmon and scrape off the marinade with a knife.

To serve 3 or 4

¾ cup	short- or round-grain white rice	175 ml.
¾ lb.	salted salmon, rinsed and sliced	350 g.
⅔ cup	milk	150 ml.
7 tbsp.	butter, 4 tbsp. [60 ml.] cut into pieces, 3 tbsp. [45 ml.] melted	105 ml.
2	eggs, the yolks separated from the whites, the yolks beaten, and the whites stiffly beaten	2
2 tsp.	salt	10 ml.
½ tsp.	white pepper	2 ml.
2 tsp.	sugar	10 ml.
3 tbsp.	dry bread crumbs	45 ml.

Bring the milk to a boil. Add the rice, and cook over medium-high heat, stirring occasionally and adding a little more milk if necessary, until the rice is almost soft—about 15 minutes. Take the rice off the heat, beat in the pieces of butter at once and let the rice cool. When it is quite cold, stir in the egg yolks, salmon, salt, pepper and sugar.

Finally, fold in the egg whites. Adjust the seasoning, and pour the mixture into a buttered 4-cup [1-liter] pie dish sprinkled with 2 tablespoons [30 ml.] of the bread crumbs. Top the mixture with the remaining bread crumbs. Bake in a preheated 350° F. [180° C.] oven for one hour, or until lightly browned. Serve with the melted butter.

ALAN DAVIDSON
NORTH ATLANTIC SEAFOOD

Neapolitan Savory Rice Mold

Sartù di Riso alla Napoletana

To serve 6 to 8

1¾ cups	short- or round-grain white rice	425 ml.
½ lb.	boneless veal, finely chopped	¼ kg.
4	eggs, beaten	4
⅔ cup	freshly grated Parmesan cheese	150 ml.
	salt and pepper	
¼ cup	flour	50 ml.
¼ cup	olive oil	50 ml.
1	onion, chopped	1
5 oz.	chicken livers and hearts, chopped	150 g.
5 oz.	sweet Italian sausage, pricked in several places	150 g.
¾ cup	puréed tomato (recipe, page 166)	175 ml.
⅓ cup	dried mushrooms, soaked in warm water for 30 minutes, drained and sliced	75 ml.
4 cups	meat stock (recipe, page 166)	1 liter
⅓ cup	shelled fresh peas	75 ml.
2½ oz.	fresh pork fat, diced	75 g.
¼ cup	dry bread crumbs	50 ml.
½ lb.	mozzarella cheese, cubed	¼ kg.

Put the veal into a bowl, and mix it with half of the beaten eggs, half of the Parmesan and a little salt and pepper. Mold the mixture into six small patties, dust them lightly with flour and fry them in the oil. When they begin to brown, add the onion, then add the livers and hearts, and the sausage. When all of the meats are well browned, add the puréed tomato and the mushrooms. Adjust the seasoning, partially cover the pan, and simmer the mixture for about 20 minutes, or until the mushrooms are tender. Remove and drain the patties and sausage.

Meanwhile, bring the stock to a boil, and add the peas and rice. Cook for about 15 minutes, or until the rice is *al dente*. Remove the pan from the heat and add ¼ cup [50 ml.] of the pork fat, the rest of the Parmesan and the rest of the beaten eggs. Stir the rice mixture well, tip it into a dish or bowl and let it cool.

Render the rest of the pork fat by lightly frying it, and use it to grease a 5-cup [1¼-liter] soufflé dish. Sprinkle the inside of the dish with a spoonful of the bread crumbs, shaking out any excess. Line the soufflé dish with about three quarters of the rice, pressing it well to make sure it sticks to the sides, and leaving a large hollow in the center. Fill this hollow with the patties; pour the chicken-giblet sauce over them. Slice the sausage and place it, and the mozzarella, on top of the patties. Cover the meat and cheese with the remaining rice, pressing it down to smooth the surface. Sprinkle it with the remaining bread crumbs and bake in a pre-heated 375° F. [190° C.] oven for about 20 minutes, or until the bread-crumb topping is golden brown. Let the *sartù* stand for about five minutes before unmolding.

IL MONDO IN CUCINA: MINESTRE, ZUPPE, RISO

Risotto with Parmesan

Risotto alla Parmigiana

The technique of making risotto is shown on pages 26-27.

	To serve 4	
2 cups	short- or round-grain white rice, preferably Italian	½ liter
1	small onion, chopped	1
6 tbsp.	butter	90 ml.
6 cups	meat stock *(recipe, page 166)*	1½ liters
⅔ cup	freshly grated Parmesan cheese	150 ml.

Fry the onion in a pan with 3 tablespoons [45 ml.] of the butter until the onion turns golden. Add the rice and, after a few minutes, add ⅔ cup [150 ml.] of the stock. Stir the mixture continuously, gradually adding more stock as it is absorbed by the rice.

When the rice is cooked—that is, when it is *al dente* and not too dry—remove the pan from the heat; add the rest of the butter, cut into pieces, and the Parmesan. Stir and serve.

STELLA DONATI
LE FAMOSE ECONOMICHE RICETTE DI PETRONILLA

Milanese Risotto

Risotto alla Milanese

	To serve 5 or 6	
2 cups	short- or round-grain white rice, preferably Italian	½ liter
1	small onion, finely chopped	1
1½ oz.	beef marrow, cut into ½-inch [1-cm.] slices	45 g.
3 tbsp.	butter	45 ml.
⅔ cup	dry white wine	150 ml.
5 cups	meat stock *(recipe, page 166)*	1¼ liters
¼ tsp.	powdered saffron, steeped in 2 tbsp. [30 ml.] hot meat stock	1 ml.
about 1 cup	freshly grated Parmesan cheese	about ¼ liter

In a shallow pan, cook the onion and the marrow in half of the butter for about four minutes, until the onion is soft. Stir

---

in the rice and, after a few minutes, add the wine, ⅔ cup [150 ml.] of the stock and the saffron. With the heat on low, stir the mixture slowly until the liquid is almost absorbed, then add a ladleful more of the stock.

Continue stirring and adding stock a little at a time. When the rice is just tender, take the pan off the heat and stir in the remaining butter and the Parmesan. Send to the table with some more Parmesan on the side.

PELLEGRINO ARTUSI
LA SCIENZA IN CUCINA E L'ARTE DI MANGIAR BENE

Chard and Squid Risotto

Risotto con Bieta e Seppie

The techniques of cleaning and preparing squid are demonstrated on page 84.

	To serve 6	
2½ cups	short- or round-grain white rice, preferably Italian	625 ml.
6 tbsp.	oil	90 ml.
6 tbsp.	finely chopped fresh parsley	90 ml.
2	garlic cloves, chopped	2
3	fresh squid, cleaned and thinly sliced	3
¾ cup	dry white wine	175 ml.
	salt and pepper	
	freshly grated nutmeg	
½ lb.	Swiss chard, stems removed and leaves coarsely chopped	¼ kg.
3	small tomatoes, peeled, seeded and chopped	3
6 cups	fish stock *(recipe, page 166)*, brought to the boiling point	1½ liters

Heat the oil in a deep pan and add the parsley and the garlic. When the garlic has browned, add the squid and cook gently for about 15 minutes, or until they become opaque.

Add the white wine. Season the mixture with salt, pepper and nutmeg. After about 10 minutes, when the wine has been absorbed in the mixture, add the Swiss chard and simmer for 10 minutes to let it absorb the flavors in the pan; then add the tomatoes.

Pour in the rice, mix well, and gradually add the stock, stirring occasionally until the rice is cooked—20 to 30 minutes. The rice should be moist and tender, but still firm.

MARIU SALVATORI DE ZULIANI
LA CUCINA DI VERSILIA E GARFAGNANA

Risotto with Peas, Venetian-Style

Risotto con Piselli alla Veneziana

The technique of making this dish is shown on page 27.

To serve 6

2 cups	short- or round-grain white rice, preferably Italian	½ liter
2 quarts	chicken or beef stock (recipes, page 166)	2 liters
8 tbsp.	butter	120 ml.
2	slices prosciutto, diced	2
1	small onion, thinly sliced	1
	salt and freshly ground pepper	
2 cups	shelled fresh peas	½ liter
1 tbsp.	finely chopped fresh parsley (optional)	15 ml.
1 cup	freshly grated Parmesan cheese	¼ liter

Bring the stock to a boil and keep it simmering. Melt half of the butter in a saucepan set over medium heat, then add the prosciutto. When the ham is very hot, add the onion. When the onion is transparent, add the rice and stir the grains until they are well coated with butter. When the rice glistens, start adding the hot stock a ladleful at a time. Allow the first ladleful to be absorbed before adding the next, but do not let the rice dry out. Season with salt and pepper and, when the rice is half-done (about 10 to 12 minutes), add the peas. Continue adding the stock and continue stirring. When the rice is almost done (about 25 minutes but, to be sure, bite one grain—it should be firm but not hard), add the parsley and Parmesan. Just before removing the pan from the heat, add the rest of the butter. Serve immediately, because the rice gets mushy after having been removed from the stove.

HEDY GIUSTI-LANHAM AND ANDREA DODI
THE CUISINE OF VENICE & SURROUNDING NORTHERN REGIONS

Rice Rolls with Spicy Filling

Lemper

The preparation of coconut milk is explained in the editor's note for Yellow Rice, Nepalese-Style, page 126. The herbal curry leaf and the trassi, a dried shrimp paste, are available at Asian food markets.

Wrapping food in banana leaves is a popular method of cooking and serving in Southeast Asia. Where banana leaves are available, prepare them by stripping the leaves from the thick middle rib with a sharp knife. The leaves are inclined to split, but this doesn't matter, for they will have to be cut into suitably sized pieces anyway. Wash the leaves to remove any dust, then pour boiling water over them. This makes them pliable enough to fold without splitting. Bamboo leaves, available at Chinese food markets, can be used instead of banana leaves, and heavy-duty aluminum foil does the job as efficiently even if not as picturesquely.

To serve 4 to 6

2 cups	glutinous rice, rinsed	½ liter
1 cup	coconut milk	¼ liter
2 tbsp.	oil	30 ml.
2	garlic cloves, crushed	2
2	curry leaves	2
2 tsp.	ground coriander	10 ml.
1 tsp.	ground cumin	5 ml.
½ tsp.	ground turmeric	2 ml.
½ tsp.	*trassi*	2 ml.
½ lb.	ground pork or finely chopped chicken	¼ kg.
about ½ tsp.	salt	about 2 ml.
about ½ tsp.	freshly ground pepper	about 2 ml.
	lemon juice	
	banana leaves or aluminum foil	

Put the rice into a saucepan with 2 cups [½ liter] of water. Bring the water to a boil, then turn the heat very low, cover the pan tightly and steam the rice for 15 minutes.

In a small saucepan, mix ¾ cup [175 ml.] of the coconut milk with ½ cup [125 ml.] of water and heat the liquid, without boiling. Add the liquid to the rice, stir it gently with a fork, cover, and steam for five to 10 minutes, or until the coconut milk has been absorbed. Set the rice aside to cool.

Heat the oil in a wok or medium-sized saucepan, and fry the garlic and curry leaves for one minute. Add the spices and the *trassi* and fry, stirring, for one minute longer, crushing the *trassi* with the back of a wooden spoon.

Season the pork or chicken with the salt and pepper, and add it to the spice mixture; continue frying until the color of the meat changes and it no longer looks raw. Add the remaining coconut milk and simmer, uncovered, over low heat until the meat is well cooked and quite dry. Add lemon juice to taste, check the seasoning, and add more salt and pepper if necessary. Let the meat cool.

To make the rice roll, take a large tablespoonful [about 15 ml.] of rice and spread it ½ inch [1 cm.] thick on a piece of prepared banana leaf. Put a good teaspoonful [about 5 ml.] of the meat filling in the middle and mold the rice around it. Roll up the banana leaf and secure with wooden picks. Heat the parcels over a barbecue grill or steam them for 15 minutes, then let them cool. They are served at room temperature as a snack.

CHARMAINE SOLOMON
THE COMPLETE ASIAN COOKBOOK

Eight-Treasure Rice Pudding
Pa-pao-fan

The technique of making this dish is demonstrated on pages 30-31. Candied angelica, made from the herb stems, is available from specialty food stores. Sweet red bean paste is available at Chinese food markets.

To serve 6

2 cups	glutinous rice, soaked overnight and drained	½ liter
¼ cup	oil	50 ml.
1½ cups	sweet red bean paste	375 ml.
12	pitted dates, each stuffed with a blanched and peeled almond	12
½ cup	candied citron, cut into squares	125 ml.
4	red candied cherries, halved	4
3	green candied cherries, halved	3
½ cup	candied dark green angelica, cut into squares	125 ml.
¾ cup	candied light green angelica, ½ cup [125 ml.] cut into slivers, the rest cut crosswise and filled with pieces of red candied cherry	175 ml.
2 tbsp.	sugar	30 ml.
Almond syrup		
¼ cup	sugar	50 ml.
1 tsp.	almond extract	5 ml.
1 tsp.	cornstarch, dissolved in 3 tbsp. [45 ml.] cold water	5 ml.

In a wok or a 10-inch [26-cm.] skillet, heat the oil over medium heat. Add the bean paste and cook, stirring frequently, until it comes away from the sides of the pan—four to five minutes. Transfer the bean paste to a small bowl and cool it.

Combine the glutinous rice with 2 cups [½ liter] of cold water, bring it to a boil, cover, and cook until small craters appear on the surface of the rice—three to four minutes. Reduce the heat and simmer the rice until soft—12 to 15 minutes. Put it into a bowl, let it cool and stir in the sugar.

With flavorless vegetable oil, coat a heatproof 2¼-quart [2¼-liter] mold that has gradually sloping sides. Line the mold with plastic wrap, letting the wrap drape over the sides all around. Oil the plastic wrap and smooth out as many wrinkles as possible. Arrange the candied fruits in a pattern that covers the bottom and extends up the sides of the mold. Then form part of the sweetened rice into a shape that will fit the decorated part of the mold, and set the rice in place over the decorations, pressing lightly. Build up the sides of the pudding with more rice and fill the center with the bean paste. Spoon the rest of the rice evenly over the bean paste, spreading the rice out to the edges of the mold. Fold the edges

of the plastic wrap over the pudding. Place the pudding on a rack in a pot filled with several inches of boiling water. Cover the pot and steam the pudding for 45 minutes, replenishing the water as necessary with more boiling water.

Make the syrup by first combining the sugar with 1 cup [¼ liter] of cold water in a small pan. Bring to a boil, stirring until the sugar dissolves. Reduce the heat to low; add the almond extract and cornstarch. Stir for a few minutes until the syrup thickens slightly and becomes clear.

Remove the pudding from the pot, peel the plastic wrap from the top and unmold the pudding onto a serving plate. Carefully peel off the remaining plastic wrap. Pour the hot syrup over the pudding and serve at once.

FOODS OF THE WORLD/THE COOKING OF CHINA

Deep-fried Glutinous Rice Rolls
Ssu Ssu Pao Chuan

The techniques of cooking glutinous rice and soybeans are explained on pages 16-17 and 30-31. Bean-curd skin is available frozen at Chinese food markets. To make brown peppercorn salt, crush equal amounts of whole black peppercorns and salt together in a mortar or pulverize them in a blender or food processor.

To serve 4

2 cups	cooked glutinous rice	½ liter
2 tbsp.	cooked soybeans	30 ml.
3	Chinese dried black mushrooms, soaked for 30 minutes in warm water, drained and diced	3
about 1¼ cups	oil	about 300 ml.
⅓ cup	diced carrots, boiled for 15 minutes	75 ml.
1 tsp.	sugar	5 ml.
¼ tsp.	pepper	1 ml.
2 tbsp.	soy sauce	30 ml.
4	sheets bean-curd skin	4
	brown peppercorn salt	

Stir fry the black mushrooms in 3 tablespoons [45 ml.] of the oil. Add the cooked soybeans and carrots, season with the sugar, pepper and soy sauce, and mix well. Stir in the cooked rice and mix thoroughly.

Cut each bean-curd sheet into a rectangle 6 inches [15 cm.] long and 4 inches [10 cm.] wide. Place about ½ cup [125 ml.] of the rice mixture on each rectangle, and wrap it into a long, thin roll. Heat the remaining oil in a pan, and fry each roll until golden brown. Cut each roll into four pieces. Arrange the pieces standing up on a plate and serve with brown peppercorn salt.

FU PEI MEI
PEI MEI'S CHINESE COOK BOOK

Sweet Brown Rice

Brown, or unshelled, sesame seeds can be purchased from health-food stores.

	To serve 6	
3 cups	glutinous rice, soaked overnight in 2 cups [½ liter] water	¾ liter
1 cup	brown sesame seeds	¼ liter
3 tbsp.	soy sauce	45 ml.

Bring the rice and soaking water to a boil; reduce the heat and simmer, covered, for one and one half hours. Transfer the rice to a large wooden bowl and pound it with a wooden pestle or heavy stick until it becomes sticky and fairly smooth. This does not take long—about 10 minutes.

Let the pounded rice cool, then wet your hands and shape the rice into lumps about the size of walnuts.

Put the sesame seeds into an ungreased, heavy skillet and roast them, stirring occasionally, until they begin to pop and darken in color. Turn off the heat, season the seeds with the soy sauce and stir the seeds to dry them. Turn the rice balls in the sesame seeds until the balls are covered all over.

TOM RIKER & RICHARD ROBERTS
THE DIRECTORY OF NATURAL & HEALTH FOODS

Savory Rice and Mushrooms

	To serve 4	
1 cup	brown rice, rinsed and drained	¼ liter
1 tsp.	salt	5 ml.
3 tbsp.	oil or butter	45 ml.
½ cup	diced onion	125 ml.
1	large green pepper, halved, seeded, deribbed and diced	1
1	pimiento, diced	1
1 cup	diced fresh mushrooms	¼ liter
2	eggs, lightly beaten	2
⅔ cup	freshly grated Parmesan cheese	150 ml.
½ cup	dry bread crumbs, sautéed in 2 tbsp. [30 ml.] butter	125 ml.

Combine the rice with 2 cups [½ liter] of boiling water and the salt in the top of a double boiler and cook over simmering water until tender—about one and one half hours. Heat the oil or butter in a skillet over medium heat and cook the onion, green pepper, pimiento and mushrooms in it for five minutes. Add the mixture to the rice along with the eggs; stir until thoroughly mixed and blend in the grated cheese. Put the mixture in a buttered 4-cup [1-liter] casserole and sprinkle the bread crumbs over it. Bake in a preheated 350° F. [180° C.] oven for 30 minutes, or until firmly set.

THE WISE ENCYCLOPEDIA OF COOKERY

Brown Rice with Currants and Scallions

	To serve 6	
1½ cups	brown rice	375 ml.
3 cups	chicken stock (recipe, page 166)	¾ liter
1 tsp.	turmeric	5 ml.
¼ cup	dried currants	50 ml.
6 to 8	scallions, finely chopped, including the green tops	6 to 8
2 tsp.	finely chopped candied ginger	10 ml.
3 tbsp.	butter, softened	45 ml.

Combine the rice, stock and turmeric in a heavy saucepan with a tight-fitting lid. Bring the stock to a boil; reduce the heat and simmer for 45 minutes, or until the rice is tender and all of the liquid has been absorbed.

Using two forks to keep the grains separated, toss the rice with the currants, scallions, ginger and butter. Let the mixture stand, covered, for a few minutes before serving.

ROBERT C. ACKART
COOKING IN A CASSEROLE

Brown Rice with Spanish Sausages

Chorizos con Arroz

	To serve 6	
2 cups	brown rice	½ liter
6	small *chorizos*	6
2 tbsp.	rendered bacon fat	30 ml.
4	slices pineapple	4
1	avocado, peeled, pitted and sliced	1
	strained fresh lime juice	
	salt	
	paprika	

Cover the *chorizos* with about 4 cups [1 liter] of water and bring it to a boil in a heavy pot. Cover the pot and simmer the *chorizos* until they are tender—20 to 30 minutes. Remove the *chorizos* and set them aside. Measure the cooking water to ensure that there is 4 cups; if not, add fresh water. Bring it to a boil, add the brown rice, cover the pot and cook the rice over low heat for 40 minutes, or until it is done and dry.

Slice the *chorizos*. Melt the bacon fat in a skillet over medium heat and brown the *chorizos* and pineapple slices. Drain them on brown paper. Mound the hot brown rice on a serving plate and surround with the *chorizos* and pineapple. Decorate with avocado slices, sprinkled with fresh lime juice, salt and paprika.

PHYLLIS JERVEY
RICE & SPICE

Wild and Brown Rice

	To serve 8	
¾ cup	wild rice, rinsed, drained and parboiled for 2 or 3 minutes	175 ml.
¾ cup	brown rice	175 ml.
4½ cups	chicken stock (recipe, page 166) or water	1,125 ml.
1½ tsp.	salt	7 ml.
	butter	
1 tbsp.	chopped fresh parsley	15 ml.

In a large saucepan with a tight-fitting lid, bring the stock or water to a boil and add the wild rice, brown rice and salt. Cover the saucepan, reduce the heat to low and simmer for 45 minutes, or until the brown rice is tender and the wild rice is well opened and curled back. Lightly butter a serving dish, put in the rice and sprinkle with the parsley.

THE JUNIOR LEAGUE OF THE CITY OF NEW YORK
NEW YORK ENTERTAINS

Wild Rice Croquettes

	To serve 4	
1 cup	wild rice, rinsed, soaked overnight and drained	¼ liter
3 cups	chicken stock (recipe, page 166)	¾ liter
1	egg, plus 4 egg yolks	1
1	thin slice of a garlic clove, chopped	1
1 tsp.	fresh onion juice	5 ml.
1 tbsp.	finely chopped fresh parsley	15 ml.
1 tbsp.	finely cut chives	15 ml.
	salt and pepper	
	Tabasco sauce	
2	chicken livers, parboiled for 2 minutes, drained and mashed with a fork	2
2 tbsp.	milk	30 ml.
	dry bread crumbs	
	oil for deep frying	

Place the rice and chicken stock in a double boiler and cook over boiling water, stirring occasionally until the liquid is absorbed—about 30 minutes.

Beat the four egg yolks with the garlic, onion, parsley, chives, salt and pepper to taste, and a dash of Tabasco sauce.

Add the egg mixture to the rice together with the livers, and cook until slightly thickened. Chill the mixture thoroughly and form it into croquettes 2 inches [5 cm.] long.

Beat the whole egg with the milk; dip the croquettes in this mixture, then roll them in the bread crumbs. Refrigerate the croquettes for at least two hours, or until ready for use. Deep fry the croquettes in oil heated to 390° F. [195° C.] until they are golden brown; drain them on absorbent paper and serve hot.

MARTHA MEADE
RECIPES FROM THE OLD SOUTH

Casserole of Carrots and Wild Rice

The technique of cooking wild rice is explained on pages 22-23. This creamy casserole of wild rice and carrots can be prepared early in the day and baked when needed. Allow an extra 10 minutes of baking time if you have kept the casserole in the refrigerator.

	To serve 6	
4 cups	cooked wild rice	1 liter
3 oz.	bacon, diced	90 g.
1	large onion, chopped	1
2 cups	finely grated carrot	½ liter
1 cup	light cream	¼ liter
1	egg, lightly beaten	1
1 tsp.	salt	5 ml.

Fry the bacon over low heat until it is crisp and brown. Remove and set aside the pieces, then sauté the onion in the bacon fat until the onion is lightly browned. Add the wild rice, the bacon and the carrot to the sautéed onion, stirring to mix well. Combine the cream with the egg and salt. Fold the egg into the rice mixture.

Turn the mixture into a buttered 2-quart [2-liter] casserole and bake, covered, in a preheated 350° F. [180° C.] oven for about 30 minutes. Remove the cover, stir the rice mixture well, and bake uncovered for another 10 minutes, or until the rice is firm.

BETH ANDERSON
WILD RICE FOR ALL SEASONS COOKBOOK

Wild Rice with Chicken Livers

To serve 4

1 cup	wild rice, rinsed, soaked overnight and drained	¼ liter
½ lb.	chicken livers	¼ kg.
8 tbsp.	butter	120 ml.
1	onion, finely chopped	1
	thyme	
	marjoram	
	salt and pepper	
	Tabasco sauce	

Cover the rice with 4 cups [1 liter] of boiling salted water; set the pan over low heat, cover it and simmer the rice until tender (from 40 to 60 minutes). Sauté the chicken livers in the butter with the onion, a pinch each of thyme, marjoram, salt and pepper, and a dash of Tabasco sauce. Drain the rice in a colander. Blend it well with the chicken-liver mixture and serve hot.

MARTHA MEADE
RECIPES FROM THE OLD SOUTH

Baked Wild Rice and Turkey

To serve 4

1 cup	wild rice, rinsed, soaked for 2 hours and drained	¼ liter
2½ cups	sliced fresh mushrooms (about ½ lb. [¼ kg.])	625 ml.
5 tbsp.	butter	75 ml.
2 cups	diced cooked turkey	½ liter
1½ cups	cream	375 ml.
2½ cups	turkey stock *(recipe, page 166)*	625 ml.
	freshly grated nutmeg	
	salt and pepper	
about ⅓ cup	shredded cheese	about 75 ml.

Sauté the mushrooms in 3 tablespoons [45 ml.] of the butter. Mix them with the turkey and the rice. Add the cream and 1½ cups [375 ml.] of the turkey stock, and season with a pinch of nutmeg, salt and pepper. Transfer the mixture to a generously buttered casserole and bake it, covered, in a preheated 350° F. [180° C.] oven for one hour. Add the remain-ing stock and bake the casserole for another 30 minutes, or until the rice is tender. Sprinkle the top with the grated cheese, dot it with the remaining butter, and brown it, uncovered, under the broiler just before serving.

NARCISSE CHAMBERLAIN AND NARCISSA G. CHAMBERLAIN
THE CHAMBERLAIN SAMPLER OF AMERICAN COOKING

Wild Rice and Oyster Casserole

The cooking of wild rice is explained on pages 22-23.

To serve 6 to 8

3 cups	cooked wild rice	¾ liter
8 tbsp.	butter	120 ml.
¾ cup	finely chopped shallots or onion	175 ml.
¾ cup	finely chopped celery	175 ml.
⅔ cup	milk	150 ml.
1 pint	shucked oysters, liquor reserved	½ liter
1 tsp.	salt	5 ml.
¼ tsp.	pepper	1 ml.
¼ tsp.	ground cardamom seeds	1 ml.
¼ lb.	Swiss cheese, shredded (about 1 cup [¼ liter])	125 g.

Heat half of the butter in a frying pan, and cook the shallots or onion and celery over low to medium heat until they are soft and barely golden. Combine them with the wild rice. Butter a 1½- or 2-quart [1½- or 2-liter] casserole and in it arrange alternate layers of rice and oysters, ending with a layer of rice.

Add enough milk to the oyster liquor to make ¾ cup [175 ml.] of liquid. Stir the salt, pepper and cardamom into the milk mixture, and pour it over the contents of the casserole. Sprinkle the top with the shredded cheese, then melt the remaining butter and sprinkle it over the cheese. Bake, covered, in a preheated 400° F. [200° C.] oven for 25 minutes, or until the cheese melts and begins to brown.

JOSÉ WILSON (EDITOR)
HOUSE & GARDEN'S PARTY MENU COOKBOOK

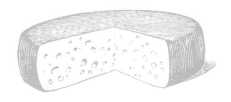

Wild Rice with Fresh Peas

The cooking of wild rice is explained on pages 22-23.

	To serve 6	
4 cups	cooked wild rice	1 liter
3 tbsp.	butter	45 ml.
2 oz.	almonds, blanched, peeled and slivered (about ½ cup [125 ml.])	60 g.
2 cups	shelled fresh peas, parboiled for 2 to 3 minutes and drained	½ liter
1 tbsp.	freshly grated orange peel	15 ml.

Melt the butter in a 2-quart [2-liter] casserole and lightly toss the almonds in the butter. Put the casserole into a preheated 325° F. [160° C.] oven and bake for 10 to 15 minutes, stirring occasionally until the almonds are lightly toasted. Add the peas, wild rice and orange peel to the casserole, and toss lightly to mix. Return the casserole to the oven and bake, uncovered, until all of the ingredients are heated through—about 15 minutes.

BETH ANDERSON
WILD RICE FOR ALL SEASONS COOKBOOK

Wild Rice Curry

The cooking of wild rice is explained on pages 22-23.

	To serve 4 to 6	
2 cups	cooked wild rice	½ liter
6 oz.	bacon, diced	175 g.
½ cup	chopped onion	125 ml.
½ cup	grated carrots	125 ml.
2	egg yolks	2
1 cup	light cream	¼ liter
1½ tsp.	curry powder	7 ml.
½ tsp.	salt	2 ml.
4 tbsp.	butter	60 ml.

Fry the bacon until crisp, add the onion and carrots, and sauté until the onions are soft. Remove the bacon, onion and carrots with a slotted spoon and mix them with the wild rice. Place the rice mixture in a buttered casserole.

Beat the egg yolks with the cream, curry powder and salt, and pour them over the rice. Dot the top of the rice with the butter. Place the casserole in a pan of boiling water and cook in a preheated 300° F. [150° C.] oven until it is set—approximately 30 minutes.

HELEN CORBITT
HELEN CORBITT'S COOKBOOK

Groats, Grits and Meals

A Famous Old Dutch Cereal

Watergruel

To make cranberry juice, boil equal amounts of fresh cranberries and water until the berries burst—five to 10 minutes; then strain off the juice through a sieve lined with four layers of cheesecloth.

	To serve 4	
1 cup	barley flakes	¼ liter
2-inch	cinnamon stick	5-cm.
1	small piece lemon peel	1
1 cup	cranberry juice	¼ liter
½ cup	raisins	125 ml.
¾ cup	sugar	175 ml.

Cook the barley flakes with the cinnamon stick and lemon peel in 3 cups [¾ liter] of water for 10 minutes. Add the cranberry juice, raisins and sugar, and cook five minutes more. Take out the lemon peel and cinnamon stick before serving the cereal.

ELIZABETH BURTON BROWN
GRAINS

Barley with Raisins and Currants

Herren-und Damen-Brei

	To serve 2 or 3	
½ cup	pearl barley	125 ml.
1 cup	raisins	¼ liter
1 cup	dried currants	¼ liter
	salt	
2	lemons, the peel grated and the juice strained	2
	sugar	

Cook the barley, raisins and currants in about 1¾ cups [425 ml.] of salted water for about 30 minutes, or until tender. Add the lemon peel and cook five more minutes. Season with lemon juice and sugar to taste.

JUTTA KÜRTZ
DAS KOCHBUCH AUS SCHLESWIG-HOLSTEIN

Barley Casserole from Mildred Oakes

To serve 4

1½ cups	pearl barley	375 ml.
11 tbsp.	butter	165 ml.
½ lb.	mushrooms, quartered	¼ kg.
2	medium-sized onions, finely chopped	2
about 3 cups	hot beef or chicken stock (recipes, page 166)	about ¾ liter
	salt	
	chopped fresh parsley	

Melt 3 tablespoons [45 ml.] of the butter in a heavy skillet and sauté the mushrooms over medium-high heat for about four minutes, until they render their juices. Transfer the mushrooms to a small dish. Add the remaining butter to the skillet and heat until the foam has subsided. Add the onions to the butter and cook them until they are just wilted. Add the barley and stir over medium-high heat until the barley has a rich golden color. Combine the barley and onions with the mushrooms in a 6-cup [1½-liter] casserole or baking dish. Add about 1½ cups [375 ml.] of hot beef or chicken stock, cover, and bake in a preheated 350° F. [180° C.] oven for 30 minutes. Add another 1½ cups of hot stock, cover again and bake for a further 30 minutes. Add salt to the casserole if necessary (the stock should be well seasoned) and additional stock if the barley has absorbed too much liquid before it is tender. Sprinkle the barley with chopped parsley before serving.

JAMES BEARD
JAMES BEARD'S AMERICAN COOKERY

Pear Gruel

Parevalling

The manner in which you may have pear gruel served to you in Denmark depends on where in the country you happen to be. It can be made with either ordinary milk or buttermilk, using either barley groats or rice. In any case, the procedure for making the gruel is the same.

To serve 4 to 6

½ cup	barley groats	125 ml.
5 cups	buttermilk	1¼ liters
1 lb.	pears (about 4)	½ kg.
⅓ cup	sugar	75 ml.

Bring the buttermilk to a boil and stir in the groats. Boil the mixture for about 45 minutes. Peel the pears and cut them into quarters. Add them, with the sugar, to the gruel. Boil the gruel for about 15 minutes more and serve.

BODIL JENSEN (EDITOR)
TAKE A SILVER DISH

Berber Couscous with Barley Grits

Cheesha Belboula

The technique of making this dish is demonstrated on pages 32-33. If you include the optional hot chilies in this recipe, remember that the volatile oils in them may make your skin sting and your eyes burn; after handling chilies, avoid touching your face and wash your hands promptly.

This unusual couscous replaces semolina with barley grits. The flavor is nutty, rustic and extraordinarily good. However, the grain is different in texture from semolina and requires three steamings. Water must be added to this kind of grain very slowly.

To serve 6 to 8

4 cups	barley grits, rinsed and drained	1 liter
8 tbsp.	unsalted butter	120 ml.
⅛ tsp.	powdered saffron	½ ml.
¼ tsp.	turmeric	1 ml.
8	sprigs each parsley and coriander, tied together	8
2 tsp.	salt	10 ml.
2 tsp.	freshly ground pepper	10 ml.
2	large tomatoes, peeled, seeded and chopped	2
1	large onion, quartered	1
2-inch	cinnamon stick	5-cm.
2 to 2½ lb.	chicken, cut into serving pieces	1 to 1¼ kg.
8	small boiling onions	8
7	young turnips, peeled and quartered	7
9	small zucchini, quartered	9
2 cups	shelled fresh peas, or dried fava or lima beans, soaked overnight, drained and skins removed	½ liter
1	small hot green or red chili, stemmed, seeded and chopped (optional)	1
2 tbsp.	butter mixed with ½ tsp. [2 ml.] each dried oregano, marjoram and thyme leaves	30 ml.
2 cups	light cream	½ liter

Melt all but 1 tablespoon [15 ml.] of the unsalted butter in the base of a *couscoussier*, without letting the butter brown. Add the saffron, turmeric, bunch of parsley and coriander, salt, pepper, half of the tomatoes, the onion and the cinnamon stick. Take the pot in your hands and give it a good swirl. Add the chicken, cover the pan and cook the mixture gently for 15 minutes. Pour in 4 cups [1 liter] of water, cover the vessel and simmer for 30 minutes more. Then add the small boiling onions.

Butter the inside of the top half of the *couscoussier*. Squeeze the barley to extract excess water and slowly add it by rubbing the grains between your palms as you drop them in. Before sealing the two halves of the *couscoussier* together with a band of cheesecloth, make sure there is plenty of liquid in the bottom pot.

Steam the barley for 20 minutes, then turn it out into a large bowl and break up the lumps with a fork. Add the reserved 1 tablespoon of unsalted butter, then begin adding water very slowly while raking and working the grains to help them swell and separate. It should take about three minutes to add 2 cups [½ liter] of water. Do not let the grains become soggy. Rake the grains and toss them, smooth them out and let them dry for 10 minutes.

Pile the barley back into the top container, reseal with cheesecloth, and steam for a further 20 minutes. Turn the grains back into the bowl and break up the lumps. Let the barley stand for about 15 minutes, working it from time to time to keep the grains separate and to prevent them from forming a lumpy mass.

Meanwhile, add the turnips to the simmering broth in the bottom half of the *couscoussier,* and cook them for 15 minutes; add a little more boiling water if the vessel is beginning to dry out.

Thirty minutes before serving, purée the remaining tomato in an electric blender or food mill. Add the purée, zucchini, peas or beans, and chili to the broth. Bring to a boil, return the barley to the top half of the *couscoussier,* reseal the two halves with cheesecloth and steam for 20 minutes.

Turn out the barley onto a large, warmed serving dish and mix in the herb butter, working it in gently with your finger tips. Smooth out any lumps with a fork. Make a well in the center of the barley and place the chicken pieces in the well. Cover them with the vegetables. Add the cream to the broth and bring it to a boil. Turn off the heat. Strain some of the broth and use it to moisten the barley. Pluck out the herbal sprigs and serve the remaining broth as a gravy.

PAULA WOLFERT
COUSCOUS AND OTHER GOOD FOOD FROM MOROCCO

Buckwheat Groats with Cheese

Kasza Hreczana ze Serem

To serve 4

1 cup	buckwheat groats, parboiled for 2 minutes and drained	¼ liter
2 cups	sour cream	½ liter
1 tbsp.	butter, melted	15 ml.
	salt and pepper	
½ lb.	farmer or pot cheese	¼ kg.
2	egg yolks	2

Combine the groats with the sour cream and the butter. Season with salt and pepper.

Put half of the mixture into the bottom of a well-buttered baking dish. Cream the cheese with the egg yolks, spread the cheese over the groats in the casserole, and then cover the cheese with the remainder of the groats. Bake, covered, in a preheated 350° F. [180° C.] oven for 50 minutes to one hour, or until the sour cream has baked to a custard consistency. Test with a sharp knife plunged into the casserole; it should emerge moist but clean. Serve with sour cream.

MARJA OCHOROWICZ-MONATOWA
POLISH COOKING

Buckwheat Groats Baked with Meat

Kasza Gryczana Zapiekana z Mięsem Lub Podrobami

This dish should be served with fresh cabbage or sauerkraut.

To serve 4

1 ½ cups	buckwheat groats	375 ml.
4 tbsp.	lard	60 ml.
⅓ cup	chopped onion	75 ml.
½ lb.	roast meat (pork, beef or lamb), finely chopped or ground	¼ kg.
¼ cup	beef stock (recipe, page 166)	50 ml.
	salt and pepper	
½ to 1 cup	heavy cream	125 to 250 ml.
1 tbsp.	mixed chopped fresh herbs such as parsley, tarragon and thyme	15 ml.

Heat half of the lard in a saucepan, stir in the buckwheat groats, then pour in about 4 cups [1 liter] of water. Bring to a boil, lower the heat and simmer the groats, covered, for 20 minutes, or until all of the water has been absorbed. Stir the groats, then cover the pan and set it aside, off the heat, until the groats are soft and dry, about 10 minutes.

Brown the onion to a light golden color in the remaining lard, then add the meat and stock. Bring the mixture to a boil, stir it into the groats, and season with salt and pepper.

Put the mixture into a buttered shallow baking dish. Pour the cream over the top, and bake in a preheated 400° F. [200° C.] oven for 20 minutes. Garnish the dish with the chopped herbs before serving.

HELENA HAWLICZKOWA
KUCHNIA POLSKA

Buckwheat Groats with Mushrooms and Onions

This dish may be cooked in advance and reheated, covered, in a preheated 200° F. [100° C.] oven for 20 minutes or so. Or, it may be steamed by placing the cooked groats in a metal colander and setting the colander over a deep pot filled with 1 inch [2½ cm.] of water. Drape the colander with a towel, bring the water to a boil, and steam the groats for about 10 minutes, or until they are heated through.

	To serve 6	
1 cup	buckwheat groats	¼ liter
1	egg	1
1 tsp.	salt	5 ml.
8 tbsp.	butter	120 ml.
3	medium-sized onions, finely chopped	3
½ lb.	mushrooms, finely chopped	¼ kg.

In a mixing bowl, toss the groats and egg together with a large wooden spoon until the grains are thoroughly coated. Transfer the mixture to an ungreased, 10- to 12-inch [25- to 30-cm.] skillet (preferably one with a nonstick surface) and cook, uncovered, over moderate heat, stirring constantly until the groats are lightly toasted and dry—about five minutes. Watch carefully for any sign of burning and regulate the heat accordingly.

Add the salt, 3 tablespoons [45 ml.] of the butter and 2 cups [½ liter] of boiling water. Stir the groats thoroughly, tightly cover the pan, and reduce the heat to low. Simmer, stirring occasionally, for about 20 minutes. If at this point the groats are not yet tender and seem dry, stir in another cup [¼ liter] of boiling water and simmer the groats 20 minutes longer, or until the water is absorbed and the grains are separate and fluffy.

Remove the pan from the heat, remove the cover, and let the groats rest undisturbed for about 10 minutes. Meanwhile, melt 3 tablespoons more of the butter in another 10- to 12-inch skillet over high heat. Add the chopped onions, lower the heat to moderate and, stirring frequently, fry for three or four minutes, or until the onions are soft and pale gold. Stir the onions into the groats and melt the remaining 2 tablespoons [30 ml.] of butter in the skillet over high heat. Drop in the mushrooms, reduce the heat to moderate, and cook two or three minutes, stirring frequently. Then raise the heat to high, and cook the mushrooms briskly until all of the liquid in the pan has evaporated.

Add the mushrooms to the groats and onions, and toss together. Taste for seasoning and serve.

FOODS OF THE WORLD/RUSSIAN COOKING

Buckwheat Groats with Almonds

	To serve 6	
2⅔ cups	buckwheat groats	650 ml.
¼ lb.	slivered almonds (about 1 cup [¼ liter])	125 g.
½ cup	olive oil	125 ml.
1	garlic clove, finely chopped	1
¼ cup	chopped onion	50 ml.
¼ cup	chopped green pepper	50 ml.
4 cups	chicken stock (recipe, page 166)	1 liter

Sauté the almonds in the olive oil until golden brown. Remove the almonds and set them aside. Add the garlic, onion and green pepper to the oil remaining in the pan, and sauté until the onion is golden brown, stirring frequently. Add the groats and the stock, and mix well.

Pour the mixture into a casserole and bake in a preheated 325° F. [170° C.] oven for 30 minutes. Stir in the almonds, and bake for a further 15 minutes, until the groats are soft and plump and all of the liquid has been absorbed.

PICTURE COOK BOOK

Bulgur Pilaf
Bulgur Pilavi

	To serve 6	
1 cup	coarse bulgur, rinsed and drained	¼ liter
5 tbsp.	butter	75 ml.
1	large onion, coarsely grated	1
1 tbsp.	puréed tomato (recipe, page 166) or 1 small tomato, peeled, seeded and diced	15 ml.
1½ cups	beef or chicken stock (recipes, page 166)	375 ml.
	salt and freshly ground pepper	

Place 2 tablespoons [30 ml.] of the butter in a heavy saucepan, add the onion, and sauté it over medium heat, stirring constantly, until it turns golden brown. Add the tomato and cook for five minutes.

Add the stock and remaining butter, and season the mixture with a little salt and pepper; bring the stock to a boil. Add the bulgur, stir once and cover the pan. Boil for five minutes over high heat, then reduce the heat and cook the bulgur until it absorbs all of the stock, which should take about 25 minutes.

Remove the pan from the heat, remove the cover, place a napkin over the saucepan and replace the cover. Leave the pan in a warm place for 40 minutes before serving.

NEŞET EREN
THE ART OF TURKISH COOKING

Shaker Indian Pudding

The authors recommend that the pudding be served with cream or with hard sauce made by creaming 8 tablespoons [120 ml.] of butter with 2 cups [½ liter] of sifted confectioners' sugar and 1 tablespoon [15 ml.] of brandy until the sauce is light and fluffy. Chill the sauce for two hours before serving.

To serve 8

¼ cup	cornmeal	50 ml.
2¼ quarts	milk	2¼ liters
¼ tsp.	freshly grated nutmeg	1 ml.
½ tsp.	ground cinnamon	2 ml.
¼ tsp.	ground ginger	1 ml.
1 tsp.	salt	5 ml.
1 tbsp.	butter	15 ml.
½ cup	molasses	125 ml.
	sugar	
2	eggs, well beaten	2
	ground allspice	

Heat five cups [1¼ liters] of the milk in a large pan, slowly sprinkle in the cornmeal, and cook for a few minutes. Remove the meal from the heat and add all of the remaining ingredients—including a pinch of allspice and the remaining 4 cups [1 liter] of cold milk. Pour the mixture into a buttered 2½-quart [2½-liter] heatproof dish; bake in a preheated 300° F. [150° C.] oven for about three hours, until a knife inserted into the pudding emerges clean.

AMY B. W. MILLER AND PERSIS W. FULLER (EDITORS)
THE BEST OF SHAKER COOKING

Settler's First Indian Pudding

For a creamier pudding, use the ingredients listed here but mix and cook them as demonstrated on pages 68-69.

To serve 8

¾ cup	yellow cornmeal	175 ml.
7 cups	milk	1¾ liters
6 tbsp.	butter	90 ml.
¼ cup	sugar	50 ml.
1 cup	molasses	¼ liter
1½ tsp.	salt	7 ml.
1¼ tsp.	grated cinnamon	6 ml.
½ tsp.	freshly grated nutmeg	2 ml.

Place 5½ cups [1,375 ml.] of the milk in a saucepan. Set the pan over medium heat and, when the milk is hot, stir in the butter, cornmeal, sugar, molasses, salt and spices. Cook, stirring constantly, for 20 minutes, or until the mixture thickens. Pour the mixture into a buttered baking dish. Carefully pour the remaining milk on top of the pudding, letting it float without stirring it in. Bake in a preheated 300° F. [150° C.] oven for three to four hours, or until set.

YVONNE YOUNG TARR
THE FARMHOUSE COOKBOOK

Cornmeal Casserole

The preparation of coconut milk is explained in the editor's note for Yellow Rice, Nepalese-Style, page 126.

To serve 6

1½ cups	cornmeal	375 ml.
1	small onion, sliced	1
2 lb.	pumpkin, peeled, seeded and sliced	1 kg.
1	carrot, sliced	1
4 tbsp.	butter	60 ml.
1½ cups	leftover cooked meat, chopped	375 ml.
1½ cups	coconut milk	375 ml.

Brown the onion, pumpkin, and carrot in the butter. Add the meat and coconut milk and bring the mixture to a boil. Simmer, covered, for 25 minutes. Stir in the cornmeal and simmer for 10 minutes longer. Pack the mixture into a buttered mold, and bake in a preheated 350° F. [180° C.] oven for 15 minutes. Remove from the mold and serve hot.

CONNIE AND ARNOLD KROCHMAL
CARIBBEAN COOKING

Corn Pone

To serve 6

1 cup	cornmeal	¼ liter
2	eggs, well beaten	2
½ cup	sugar	125 ml.
½ tsp.	salt	2 ml.
½ cup	lard	125 ml.
1 cup	buttermilk	¼ liter
1 tsp.	baking soda	5 ml.
1 cup	flour	¼ liter
	half-and-half cream	

Mixing thoroughly after each addition, combine in order the eggs, sugar, salt, lard, buttermilk, baking soda, cornmeal and flour. Turn the mixture into a well-greased 8-inch [20-cm.] square cake pan, and bake in a preheated 350° F. [180° C.] oven for 30 minutes, or until a wooden pick plunged in the center comes out clean. Serve square cuts of the corn pone in soup dishes with half-and-half cream poured over.

EDNA EBY HELLER
THE ART OF PENNSYLVANIA DUTCH COOKING

Cornmeal Puffs

These puffs, cooled after baking, may be filled with chicken, tuna, shrimp or crab salad or cocktail spreads to serve as luncheon entrées or party snacks.

To make 10 to 12 large puffs, or 2½ to 3 dozen cocktail-sized puffs

½ cup	cornmeal	125 ml.
8 tbsp.	butter	120 ml.
¼ tsp.	salt	1 ml.
¾ cup	flour	175 ml.
4	eggs	4

Combine the butter and salt with 1 cup [¼ liter] of water in a medium-sized saucepan and bring to a boil. Add the cornmeal and flour all at once, stirring rapidly over medium heat, and cook the mixture until it leaves the sides of the pan and forms a compact ball. Continue cooking for three minutes longer, mashing the dough against the sides of the pan.

Remove the pan from the heat and, to cool the dough slightly, either turn it into the small bowl of an electric mixer and beat it for about one minute, or beat it with a sturdy wooden spoon for about two minutes.

Beat the eggs into the dough, one at a time. Continue beating until the dough is smooth and has a satiny look.

Drop the dough by tablespoonfuls onto an ungreased baking sheet, swirling the tops to round off the puffs. Space large puffs 2 inches [5 cm.] apart, small puffs 1½ inches [4 cm.] apart. Bake in a preheated 375° F. [190° C.] oven. After 50 minutes, slit each puff with a knife point and return to the oven for 10 minutes to crisp the centers.

JEANNE A. VOLTZ
THE FLAVOR OF THE SOUTH

Indian Cheese Soufflé

To serve 4 to 6

⅔ cup	cornmeal	150 ml.
4 cups	milk	1 liter
1 tsp.	salt	5 ml.
4 tbsp.	butter	60 ml.
6 oz.	Cheddar cheese, shredded (about 1½ cups [375 ml.])	175 g.
6	eggs, 1 whole, the remaining yolks separated from the whites, and the whites stiffly beaten	6

Pour the milk and salt into a saucepan and bring it to a boil; sprinkle the cornmeal over the milk and pour the mixture into a buttered baking dish. Bake in a preheated 350° F. [180° C.] oven for 20 minutes.

Remove the cornmeal mixture from the oven and place it in a mixing bowl; add the butter, cheese, egg yolks and whole egg, and stir the mixture until well blended. Gently fold in the egg whites, and pour the mixture into a buttered 4-cup [1-liter] soufflé dish. Place the soufflé in a pan of hot water, and bake in the 350° F. oven for 45 minutes, or until the soufflé is puffed and golden.

CAROL G. EMERLING AND EUGENE O. JONCKERS
THE ALLERGY COOKBOOK

Bacon Spoon Bread

To make one 8-inch [20-cm.] round bread

¾ cup	fine yellow cornmeal	175 ml.
½ lb.	sharp Cheddar cheese, coarsely shredded (about 2 cups [½ liter])	¼ kg.
4 tbsp.	butter	60 ml.
2	garlic cloves, crushed	2
½ tsp.	salt	2 ml.
1 cup	milk	¼ liter
4	eggs, the yolks separated from the whites, the yolks beaten, and the whites stiffly beaten	4
½ lb.	bacon, fried, drained and crumbled	¼ kg.

Stir the cornmeal into 1½ cups [375 ml.] of cold water in a saucepan, and place the pan over medium heat. Bring the mixture to a bubbling boil, stirring constantly. When the cornmeal is cooked to a thick batter—in perhaps 60 seconds—remove the pan from the heat. Stir in the cheese, butter, garlic and salt. When the cheese melts, stir in the milk, then the egg yolks and bacon bits. Fold the beaten egg whites into the batter.

Pour the batter into a buttered 2-quart [2-liter] casserole or soufflé dish. Level the batter with a rubber spatula and bake in a preheated 325° F. [160° C.] oven for one hour, or until a knife inserted into the center comes out clean and dry. Serve hot.

BERNARD CLAYTON JR.
THE COMPLETE BOOK OF BREADS

Baked Polenta with Fontina Cheese

Polenta al Forno con Fontina

To serve 6

2¼ quarts	cooked polenta (recipe, page 165)	2¼ liters
½ to 1 lb.	Fontina cheese, shredded	¼ to ½ kg.
⅓ cup	freshly grated Parmesan cheese	75 ml.
6 tbsp.	butter	90 ml.

Wet a marble-topped table or a large cookie sheet with cold water and pour the polenta onto it. Flatten it with the blade of a wet knife or a spatula until it is about ¼ inch [6 mm.] thick, and let it cool completely.

Butter a rectangular baking dish, making sure that the sides and bottom are well coated. Cut a sheet of polenta and place it on the bottom of the baking dish. Cover the polenta with Fontina. Top the cheese with a layer of polenta, then a layer of cheese, then polenta, until all of the polenta and Fontina have been used up. The last layer should be one of polenta. Sprinkle the Parmesan on top, dot with the butter or melt the butter and sprinkle it on, and bake in a preheated 400° F. [200° C.] oven for about 30 minutes, or until the top is golden.

HEDY GIUSTI-LANHAM AND ANDREA DODI
THE CUISINE OF VENICE & SURROUNDING NORTHERN REGIONS

Polenta with Mushrooms

Polenta con Funghi

To serve 6

6 cups	cooked polenta (recipe, page 165)	1½ liters
6 tbsp.	butter	90 ml.
¼ cup	flour	50 ml.
2 cups	milk, brought to boiling point	½ liter
½ tsp.	salt	2 ml.
¼ tsp.	white pepper	1 ml.
⅛ tsp.	freshly grated nutmeg	½ ml.
½ cup	freshly grated Parmesan cheese	125 ml.
1 lb.	fresh mushrooms, thinly sliced	½ kg.
⅓ cup	dried wild mushrooms (such as cepes), soaked in warm water for 15 minutes, drained and chopped	75 ml.

Spread the hot polenta 2 inches [5 cm.] thick on a wooden board to cool for about one hour. Meanwhile, in a small saucepan, melt 4 tablespoons [60 ml.] of the butter, blend in the flour, and cook over low heat for about three minutes, stirring constantly with a wooden spoon, until the mixture is smooth. Remove the pan from the heat and stir in the milk a little at a time. Return the pan to the heat, and continue to cook and stir the sauce for a few minutes until it is creamy and smooth. Add the salt, pepper, nutmeg and Parmesan.

In a second saucepan, cook the fresh mushrooms with the remaining butter over moderate heat for five minutes, until the mushrooms are soft. Stir in the dried mushrooms and continue cooking for five minutes more.

Slice the cooled, firmed polenta into thin strips; cover the bottom of a buttered 9-inch [23-cm.] pie dish with a layer of the strips. Spoon on a layer of sauce, spread the mushroom mixture over the sauce, then add another layer of polenta strips. Continue layering the sauce, mushrooms and polenta strips, until all of the ingredients are used; the top layer should be sauce. Bake in a preheated 425° F. [220° C.] oven for 15 minutes, and serve hot.

TERESA GILARDI CANDLER
THE NORTHERN ITALIAN COOKBOOK

Hominy Drop Cakes

The cooking of hominy is explained on pages 14-15.

Slightly sweetened, these may be served as hot cakes at breakfast with ham or sausage.

To serve 6 to 8

2 cups	cooked cracked hominy	½ liter
1 tbsp.	milk or water	15 ml.
2	eggs, the yolks separated from the whites, the whites stiffly beaten	2
2 tbsp.	flour	30 ml.
½ tsp.	salt	2 ml.

Put the hominy and the milk or water in the top of a double boiler to heat for five minutes over hot water. Gently break the hominy apart with a fork, taking care not to mash the large grains. Beat the egg yolks with the flour and the salt. Remove the hominy from the heat and blend it with the yolk mixture. There should be just enough heat in the hominy to thicken the yolks slightly. Fold the beaten egg whites into the hominy mixture. Drop by tablespoonfuls onto a heated, buttered baking sheet and slip into a preheated 400° F. [200° C.] oven for five minutes to brown.

MORTON G. CLARK
FRENCH-AMERICAN COOKING

Garlic Grits

To serve 12 to 16

2 cups	hominy grits	½ liter
6 oz.	Cheddar cheese, shredded (about 1½ cups [375 ml.])	175 g.
3	garlic cloves, crushed	3
8 tbsp.	butter	120 ml.
3	eggs, lightly beaten	3
2 cups	milk	½ liter
¼ cup	freshly grated Parmesan cheese	50 ml.

Cook the grits in 2 quarts [2 liters] of boiling salted water until they are tender but still pourable. Remove the grits from the heat; add the Cheddar, garlic and butter, and stir until the cheese melts. Let the mixture cool, then add the eggs and milk. Pour the grits into two buttered 2-quart [2-liter] casseroles. Bake in a preheated 325° F. [160° C.] oven for 50 to 60 minutes. Remove the casseroles from the oven, sprinkle the grits with the Parmesan, and bake for about 10 more minutes. Serve immediately.

THE JUNIOR LEAGUE OF NEW ORLEANS
THE PLANTATION COOKBOOK

Turkey-Hominy Croquettes

The cooking of grits is explained on page 34. The authors suggest that leftover ham or chicken may replace the turkey.

To serve 4

1 cup	cooked hominy grits	¼ liter
1 tbsp.	finely chopped onion	15 ml.
2 tbsp. plus 2 tsp.	butter, melted	40 ml.
½ cup	milk	125 ml.
1 cup	finely chopped cold roast turkey	¼ liter
2 tbsp.	chopped fresh parsley	30 ml.
	salt and freshly ground pepper	
2	eggs, beaten separately	2
	dry bread crumbs	
	oil or fat for deep frying	

In a saucepan, sauté the onion in 2 teaspoons [10 ml.] of the butter until it is soft. Add the milk, and stir in the hominy grits and turkey, the remaining butter and the parsley; season with salt and a generous grinding of pepper. Stir the mixture well and, when it is heated through, remove it from the heat and add one beaten egg. Cook for another minute, stirring constantly. Spread the mixture on a platter to cool.

When the mixture is firm, shape it into eight to 10 cylindrical croquettes, each about 3 inches [7½ cm.] long. Roll these croquettes in the other beaten egg, then in bread crumbs, and deep fry them in oil or fat heated to 375° F. [190° C.] until they are golden brown.

NARCISSE AND NARCISSA G. CHAMBERLAIN
THE CHAMBERLAIN SAMPLER OF AMERICAN COOKING

Cheese-fried Grits

To serve 4

1 cup	hominy grits	¼ liter
	salt	
	cayenne pepper	
1 cup	shredded sharp Cheddar cheese	¼ liter
1	egg, well beaten	1
	rendered bacon or ham fat, melted	

Bring 5 cups [1¼ liters] of water to a boil and slowly add the grits. Stir in salt and cayenne. Cook slowly for 20 to 30 minutes, until very thick. Let the grits cool, add the cheese and mold the mixture into croquettes. Refrigerate the croquettes for several hours or overnight. Dip them in the egg and fry them in melted fat for about five minutes, turning them frequently, until they are crisp and brown on all sides.

EDGAR THARP AND ROBERT E. JAYCOXE
THE STARVING ARTIST'S COOKBOOK

Hominy and Cheese Timbales, Russian-Style

The cooking of hominy is explained on pages 14-15.

To serve 6

2 cups	cooked cracked hominy, drained	½ liter
2	eggs, plus 1 egg yolk, beaten	2
⅔ tsp.	salt	3 ml.
⅔ cup	shredded Gruyère cheese	150 ml.
	pepper	
1 tsp.	paprika	5 ml.
2 tsp.	finely chopped green pepper	10 ml.
2 tsp.	finely chopped red pimiento	10 ml.
2 tsp.	finely chopped fresh parsley	10 ml.
1 cup	half-and-half cream, scalded	¼ liter

Beat the eggs and egg yolk with the salt. Combine the hominy, cheese, beaten eggs, a pinch of pepper, the paprika, green pepper, pimiento and parsley. Stir in the half-and-half cream, mixing thoroughly.

Pour the mixture into six generously buttered 1-cup [¼-liter] ramekins; set these in a pan of hot water and bake in a preheated 300° to 325° F. [150° to 160° C.] oven for 30 minutes, or until firmly set.

LOUIS P. DE GOUY
THE GOLD COOK BOOK

Millet Soufflé

The technique of cooking millet is explained on pages 14-15.

To serve 4 to 6

2 cups	cooked millet	½ liter
1 tsp.	salt	5 ml.
¾ cup	milk	175 ml.
½ cup	shredded Cheddar cheese	125 ml.
3	eggs, the yolks separated from the whites, the yolks beaten and the whites stiffly beaten	3

Stir together the millet, salt, milk, most of the cheese, and the egg yolks. Fold the egg whites into the mixture and pour it into an ungreased 6-cup [1½-liter] soufflé dish. Sprinkle the remaining cheese on top. Bake in a preheated 375° F. [190° C.] oven for about 30 minutes, or until a knife inserted in the center comes out clean. (Reduce the oven temperature to 325° F. [160° C.] after 20 minutes if the soufflé seems to be browning too fast.) Serve immediately.

MARLENE ANNE BUMGARNER
THE BOOK OF WHOLE GRAINS

Millet with Nuts and Raisins

Millet Judea

The author of this recipe suggests that the dish be served with poultry or game.

	To serve 6	
1 cup	millet	¼ liter
3 cups	chicken stock *(recipe, page 166)*	¾ liter
1	large onion, chopped	1
8 tbsp.	butter	120 ml.
1 tsp.	salt	5 ml.
¼ cup	golden raisins	50 ml.
1 oz.	pistachios or almonds, toasted and slivered (about ¼ cup [50 ml.])	30 g.

Stirring constantly, toast the millet in an ungreased skillet over medium heat for three to four minutes, or until golden. Add all of the remaining ingredients except the nuts. Tightly cover the skillet, reduce the heat to low, and simmer the mixture for 15 minutes, or until the millet is tender and the stock is absorbed. Stir in the nuts and toss lightly. Correct the seasoning and serve.

EILEEN GADEN
BIBLICAL GARDEN COOKERY

Tchoomak Hash with Mushrooms

Tchoomak is the Ukrainian word for a carter or wagoner. This hearty dish was supposed to see him through long journeys over difficult and often dangerous roads.

	To serve 4	
2 cups	millet	½ liter
2 to 2½ lb.	mushrooms, coarsely chopped	1 to 1¼ kg.
¼ lb.	lean bacon or salt pork with the rind removed, parboiled for 3 minutes, drained and diced	125 g.
3 tbsp.	butter or lard	45 ml.
1½ cups	chopped onions	375 ml.
	salt	
1½ tsp.	finely chopped fresh mint leaves	7 ml.

Bring 4 cups [1 liter] of water to a boil and add the millet. Simmer until half-done—about 15 minutes.

Fry the mushrooms with the bacon or salt pork in the butter or lard over medium heat until half-cooked—about two minutes. Add the onions and cook for one more minute. Drain off the excess fat. Stir the mushrooms, bacon and onions into the millet and season with salt. Place the mixture in a buttered shallow 8-inch [20-cm.] baking dish. Bake uncovered in a preheated 350° F. [180° C.] oven for three quarters to one hour, or until the hash is dry and crisp. Before serving, garnish with finely chopped mint.

N. I. GEORGIEVSKY, M. E. MELMAN,
E. A. SHADURA AND A. S. SHEMJAKINSKY
UKRAINIAN CUISINE

Coconut Millet Pudding

The preparation of coconut milk is explained in the editor's note for Yellow Rice, Nepalese-Style, page 126.

	To serve 6	
¼ cup	millet meal	50 ml.
2 cups	milk, scalded	½ liter
½ cup	sugar	125 ml.
1 cup	freshly grated coconut	¼ liter
1 tsp.	salt	5 ml.
1 tsp.	vanilla extract	5 ml.
2 tbsp.	fruit jam or preserves	30 ml.

Mix the millet meal, milk and sugar in the top of a double boiler, and cook the mixture over boiling water for 15 minutes, stirring occasionally. Stir in the coconut and salt and cook 10 more minutes. Add the vanilla.

Pour the pudding mixture into six custard cups and chill. Place 1 teaspoon [5 ml.] of jam or preserves in the center of each pudding before serving.

ELIZABETH BURTON BROWN
GRAINS

Fried Millet

Hirsotto

For a lighter taste, the author suggests cooking the onions in fat or oil instead of butter.

	To serve 4	
1 cup	millet	¼ liter
2	onions, chopped	2
6 tbsp.	butter	90 ml.
3½ cups	meat or vegetable stock *(recipes, page 166)*	875 ml.
½ cup	shredded cheese, such as Gruyère	125 ml.

Heat 3 tablespoons [45 ml.] of the butter in a skillet and sauté the onions until they are half-cooked. Add the millet and fry for about three minutes, stirring constantly. Heat the stock, pour it over the millet and cook, stirring frequently, for 14 to 18 minutes, or until the millet is tender. Fold in the remaining butter and the cheese. Cover the mixture and let it stand for a few minutes before serving.

EVA MARIA BORER
TANTE HEIDI'S SWISS KITCHEN

Fermented Millet Pancakes

Injera

Injera is the national bread of Ethiopia; it is made on large, flat, griddle-like pans placed directly over a fire. The batter is usually made from *teff*, a particularly fine flour made from millet, and is left to ferment for several days, leavened with starter kept from a previous batch. It is poured onto the griddle in a large spiral, starting at the outside and working in, and results in a thin, flexible pancake from 9 to 13 inches [23 to 32 cm.] in diameter. The pancakes are served with nearly every meal in Ethiopia, piled in a bowl or on a plate, and we have found them to be particularly good with soups and stews. Tear off a small piece, as the Southwesterners do with flour tortillas, and use it to scoop up your food.

To make 12 to 15 pancakes

3 cups	millet flour	¾ liter
¼ oz.	package active dry yeast	7½ g.
1 tsp.	honey	5 ml.
¼ tsp.	baking soda	1 ml.

Dissolve the yeast in ¼ cup [50 ml.] of lukewarm water and stir in the honey. When the yeast is frothy and rising actively—after about five minutes—dilute it with a little less than 5 cups [1¼ liters] more lukewarm water. Stir in the millet flour, blending until the lumps are gone, and set the batter aside to ferment for 24 hours. Just before cooking the pancakes, stir the batter vigorously and fold in the baking soda, making sure you eliminate any lumps.

Heat a large, well-seasoned skillet over medium heat until a small spoonful of batter cooks quickly without browning; when this test indicates that the pan is heated, pour about ⅓ cup [75 ml.] of batter onto it in a spiral, beginning at the outside of the pan and working quickly into the center. Cover the pan and cook the pancake for about one minute. The pancake should rise slightly and, when cooked, can be removed easily with a long spatula: The top should be slightly moist, the bottom dry but not crisp or brown.

Lift the cooked pancake onto a platter to cool, then stack the cooled pancakes on a serving plate. Continue cooking and cooling the pancakes until all of the batter is used.

MARLENE ANNE BUMGARNER
THE BOOK OF WHOLE GRAINS

Indian Millet

The technique of making millet pilaf is demonstrated on pages 24-25. Millet pilafs are open to a wide range of variations: Potato or another firm vegetable may be used in place of, or along with, the cauliflower; green or red peppers may be included; and the number of onions and the mixture of spices may be varied practically without limit.

To serve 6

1⅓ cups	millet	325 ml.
4 tbsp.	butter	60 ml.
2	onions, 1 thinly sliced, 1 chopped	2
2	zucchini, sliced	2
½	cauliflower, separated into florets	½
4	cardamom pods	4
1 tsp.	cumin seeds	5 ml.
4	whole cloves	4
2	garlic cloves, chopped	2
	salt and freshly ground pepper	

Heat half of the butter in a large skillet and fry the onion rings over medium heat, without stirring, until they are dry and a nice, deep cinnamon brown. Remove them to a plate and keep them in a warm place.

In the same skillet, sauté the chopped onions, zucchini and cauliflower in the remaining butter until they are lightly browned. Add the millet, spices and garlic, and stir everything around for three to five minutes. Add enough water to cover the ingredients, season with salt and pepper, and simmer—stirring and forking up the millet from time to time to help keep it light and separated—for about 20 minutes, or until the millet is tender.

Remove the skillet from the heat, cover it tightly and let the millet steam for five minutes. Fork up the grains again and serve with the onion rings scattered over the top.

TERENCE AND CAROLINE CONRAN
THE COOK BOOK

Veracruz Tamales

Tamales Estilo Veracruzano

The techniques of making tamales are demonstrated on pages 36-37. You may substitute dried cornhusks or aluminum foil for the banana leaves specified in this recipe. Anchos are large, reddish-brown chilies that range from mild to hot in taste. Both the cornhusks and chilies are sold at Latin American food markets. The volatile oils in hot chilies may make your skin sting and your eyes burn; after handling chilies, avoid touching your face and wash your hands promptly.

These tamales, typical of Veracruz, Mexico, are flavored by avocado leaves and by the banana leaves in which they are cooked. They freeze extremely well covered with foil. To reheat, put the foil packets into a preheated 350° F. [180° C.] oven; warm for about 40 minutes.

	To make 20 tamales	
2 cups	masa harina	½ liter
1 lb.	boned pork Boston shoulder, fat included, cut into small cubes	½ kg.
¼	onion	¼
2	garlic cloves	2
2 tsp.	salt	10 ml.
3	*ancho* chilies	3
1	medium-sized tomato	1
9½ tbsp.	lard, at room temperature	142 ml.
20	pieces banana leaves, each about 7 by 9 inches [18 by 23 cm.]	20
3	avocado leaves, broken into pieces	3

Put the pork, onion, one garlic clove and ½ teaspoon [2 ml.] of the salt into a saucepan. Cover the ingredients with water and bring it to a boil. Reduce the heat and simmer the pork for about 35 minutes. Let the pork cool off in the broth, then strain the meat, reserving the broth.

Heat a griddle or skillet and toast the chilies lightly over medium heat, turning them from time to time so that they do not burn. While the chilies are soft and pliable, remove the seeds and veins, then cover the chilies with hot water and let them soak for about 15 minutes. Remove them with a slotted spoon and put them into the jar of a blender. Broil the tomato until soft and blend it—peel, seeds and flesh—to a smooth sauce with the chilies, the remaining garlic clove, ½ teaspoon of the salt and ⅓ cup [75 ml.] of the pork broth.

Melt 1½ tablespoons [22 ml.] of the lard in a saucepan and, when the lard is hot but not smoking, cook the chili sauce in it for about five minutes, stirring all the time. Add the pork and ⅓ cup of the broth, and cook for about five minutes over medium heat until the liquid has reduced a little. Set the mixture aside. Beat the remaining lard for five minutes. Mix the masa harina to a dough with 1¼ cups [300 ml.] of cold water and the remaining salt. Beat ½ cup [125 ml.] of the broth and the dough alternately into the lard, then beat the resulting dough for about five minutes.

Pass the banana leaves quickly over a flame to make them more flexible. On each leaf spread a large spoonful of the dough ⅛ to ¼ inch thick [3 to 6 mm.] over an area about 3 by 4 inches [8 by 10 cm.]. Put two cubes of the meat and a little of the sauce into the center of the dough, then add a small piece of avocado leaf. Fold the edges of the banana leaf over until they cover the dough and filling. Stack the tamales in a steamer lined with additional banana leaves. Cover the tamales with more banana leaves, and then cover the top of the steamer with a thick cloth or piece of toweling. Steam the tamales for about one hour, until the dough pulls away easily from the banana leaves and is completely smooth.

DIANA KENNEDY
THE CUISINES OF MEXICO

Tamales

The techniques of making and filling tamales are demonstrated on pages 36-37.

Buy the dried cornhusks in a Mexican food market, or prepare your own by drying fresh husks in the sun or in an oven on very low heat. Trim the husks for tamales by cutting off the tops and bottoms of the leaves to make a flat leaf with squared ends. These tamales may be frozen and then reheated in a covered casserole in a moderate oven. They may also be reheated from room temperature by toasting on a griddle or in a frying pan.

	To make 36 tamales	
4 cups	masa harina	1 liter
about 40	dried cornhusks	about 40
1 cup	lard	¼ liter
½ tsp.	baking powder	2 ml.
1½ tsp.	salt	7 ml.
1½ cups	chicken stock (recipe, page 166)	375 ml.

Soak the cornhusks in hot water for three hours or more; drain the husks and dry them well on absorbent paper. Tear some of the husks into 36 narrow strips to be used in tying the tamales closed. Set the husks aside.

To make the tamales, beat the lard in a large bowl with a whisk or an electric mixer for five minutes, or until the lard is light and spongy.

Mix the baking powder and salt into the masa harina; gradually beat 1 cup [¼ liter] of the dry ingredients into the lard. Heat the stock until lukewarm and add ½ cup [125 ml.] of it very slowly to the dough, beating constantly. Continue adding cups of the dry mixture alternately with half cups of stock, beating constantly, until both are completely mixed into the lard. Beat the dough three to four minutes longer. To test for doneness, roll a small amount of the tamale dough into a ball and place it gently in a glass of cold water. When the ball of dough floats, the dough is ready.

Spread a spoonful of dough evenly in a rectangle about ⅛ inch [3 mm.] thick in the center of the inside of each husk, slightly closer to the broad end rather than the pointed end. If the husk is too narrow, use two overlapping husks. Leave enough empty space at the sides and the ends of the husk to fold over the dough—an area about twice the size of the dough rectangle. Spread a small spoonful of cheese or another filling down the center of the dough rectangle. Fold in the sides of the husk, fold over the pointed end, and then fold over the broad end. Tie a strip of cornhusk across the top flap to close the tamale. Loosely stack the tamales seam side down in a steamer that has been lined with additional husks. Cover tightly and steam for one hour. To test for doneness, open one tamale; the dough should be light in texture and cooked through.

ANGELES DE LA ROSA AND C. GANDIA DE FERNANDEZ
FLAVORS OF MEXICO

Oat or Barley Dumplings

To serve 4

1 cup	rolled oats or barley flakes	¼ liter
8 tbsp.	butter	120 ml.
3	eggs	3
	salt	
	meat or vegetable stock *(recipes, page 166)*	

Beat the butter until it is creamy, and gradually stir in the eggs, the oats or barley, and salt to taste. Drop spoonfuls of the dumpling batter into boiling meat or vegetable stock, and simmer the dumplings for 15 minutes. Serve the dumplings hot with the stock.

PEGGIE BENTON
FINNISH FOOD FOR YOUR TABLE

Down-South Raisin Rye

This may be served as an accompaniment to chicken or ham. If served as a main dish, a cupful or more of diced cooked ham may be added for the last five minutes of cooking.

To serve 4

2 cups	cracked rye	½ liter
2	smoked ham hocks	2
½ cup	raisins	125 ml.
	salt	

Put the ham hocks, raisins and 4 cups [1 liter] of water in a heavy pan and bring to a boil. Sprinkle in the cracked rye. When the water has returned to a boil, cover the pan, reduce the heat to low and simmer for 45 minutes, or until the rye is tender. Taste for seasoning and add salt if necessary.

ELIZABETH ALSTON
THE BEST OF NATURAL EATING AROUND THE WORLD

Semolina with Tomato Sauce

Semoule de Ble Dur Sauce Tomate

To serve 6

⅔ cup	semolina	150 ml.
	salt	
2	eggs	2
4 tbsp.	butter	60 ml.
1 cup	shredded Gruyère cheese	¼ liter
2 cups	tomato sauce *(recipe, page 167)*	½ liter

Bring 4½ cups [1,125 ml.] of water to a boil, add a pinch of salt and sprinkle in the semolina, stirring constantly with a wooden spoon. Cook for about 15 minutes over low to medium heat, stirring to prevent the semolina from sticking to the pan. Remove the pan from the heat and beat in the eggs, butter and half of the cheese. Season with pepper to taste and more salt if needed. Spread the cooked semolina on a flat plate or marble surface and leave it to cool. When the cooked semolina is firm and thoroughly cooled, cut it into 2-inch [5-cm.] squares. Arrange the squares in an oiled ovenproof dish and cover them with the tomato sauce. Sprinkle the remaining cheese on top and bake in a preheated 400° F. [200° C.] oven for 15 to 20 minutes, or until the sauce is bubbly and the cheese is well melted.

CHRISTIANE SCHAPIRA
LA CUISINE CORSE

Sweet Fried Semolina

Frittura Dolce

To serve 6 to 8

1 cup	semolina or farina	¼ liter
4 cups	milk	1 liter
2	eggs, beaten, plus 3 yolks	2
½ tsp.	salt	2 ml.
¼ tsp.	sugar	1 ml.
2 tsp.	grated lemon peel	10 ml.
2 tsp.	almond extract	10 ml.
½ cup	flour	125 ml.
1 cup	dry bread crumbs	¼ liter
6 tbsp.	butter	90 ml.
6 tbsp.	oil	90 ml.

In a saucepan, heat the milk just to boiling; lower the heat and add the semolina or farina a little at a time, stirring constantly with a wooden spoon. Continue cooking and stirring for 10 minutes longer.

Remove the mixture from the heat and rapidly mix in the egg yolks, salt, sugar, lemon peel and almond extract. Oil a flat surface and spread the mixture 1 inch [2½ cm.] thick to cool. The mixture should be made up to this point and refrigerated the day before you wish to use it; it is easier to cut when chilled.

The next day, cut the chilled mass crosswise into diamonds about 2 inches [5 cm.] long. Dredge the diamonds in flour, shaking off any excess. Dip them into the eggs and dredge them in bread crumbs, again shaking off any excess.

In a large skillet melt the butter over medium heat; add the oil. Fry the coated diamonds about three minutes on each side until they are golden brown. Remove and place on paper towels to drain. Serve hot.

TERESA GILARDI CANDLER
THE NORTHERN ITALIAN COOKBOOK

Moroccan Wheat Casserole

Orissa

To caramelize sugar, dissolve it in a quarter of its volume of water, then set the pan over medium heat. Bring the mixture to a boil, stirring constantly. Then stop stirring and let the syrup boil just until it begins to brown.

The quantity of wheat may be increased or decreased according to taste, as long as two parts of water are used to one part of wheat.

To serve 6 to 8

3 cups	wheat berries	¾ liter
1 cup	oil	¼ liter
1 tbsp.	chopped red pepper	15 ml.
1 tbsp.	paprika	15 ml.
2	garlic cloves	2
1 tbsp.	caramelized sugar	15 ml.
2 tbsp.	salt	30 ml.
1 lb.	beef brisket	½ kg.
6 to 8	eggs	6 to 8
about ¼ cup	flour, mixed with about 2 tbsp. [30 ml.] water and a few drops of oil to make a thick paste	about 50 ml.

Place all of the ingredients—including the eggs in their shells but not including the flour-and-water paste—in a large, deep pot with a tight-fitting lid. Add 6 cups [1½ liters] of water. Roll the paste into a long, narrow strip and press it to the rim of the pot to seal the lid. Cook the stew in a preheated 275° F. [140° C.] oven for seven hours or overnight.

IRENE F. DAY
KITCHEN IN THE KASBAH

Wheat with Meat

Soweeta

The volatile oils in hot chilies may make your skin sting and your eyes burn; after handling chilies, avoid touching your face and wash your hands promptly.

This dish should be served with a green vegetable or salad; eaten with yogurt, it makes a complete, filling meal.

To serve 4

2 cups	wheat berries	½ liter
½ cup	peeled and grated fresh ginger	125 ml.
¼ cup	ground coriander seeds	50 ml.
2 tsp.	salt	10 ml.
8 tbsp.	clarified butter	120 ml.
5	onions, sliced	5
1½ lb.	lean boneless lamb leg or shoulder, cut into 2-inch [5-cm.] cubes	¾ kg.
1 cup	yogurt, beaten with a fork until smooth	¼ liter
3 cups	milk	¾ liter
3 to 6	dried hot red chilies, about 4 inches [10 cm.] long, stemmed, seeded and soaked for 20 minutes in 1 cup [¼ liter] hot water	3 to 6

Place the fresh ginger, coriander and salt in a blender with 6 tablespoons [90 ml.] of water. Reduce the ingredients to a fine paste. Remove the paste from the blender, and set it aside on a plate near a warm place.

In a medium-sized fireproof casserole, melt 4 tablespoons [60 ml.] of the clarified butter and fry ½ cup [125 ml.] of the onions until they are just crisp. Add the lamb. Stir and immediately add the ginger mixture. Fry the meat, turning it with a spatula from time to time, for five minutes over medium heat. Add 2 cups [½ liter] of water and simmer, uncovered, until the liquid is nearly absorbed. Add the yogurt and the remaining onions. Stir and continue to simmer until the yogurt is nearly absorbed.

While the meat is cooking, bring 3 cups [¾ liter] of lightly salted water to a boil in a saucepan. Stir in the wheat berries and boil them, uncovered, until all of the water is absorbed—about 30 minutes.

When the meat and wheat are done, stir them together in the casserole. Heat the milk in a saucepan, pour it down the sides of the casserole and bring it to a boil. Cover the casserole tightly with foil, crimping the edges to form a seal. Set a lid over the foil and bake in a preheated 350° F. [180° C.] oven for 30 minutes, or until the liquid is absorbed.

Drain the chilies. Season them with salt, and sauté them gently in the remaining butter for five minutes. Pass the mixture separately as a sauce for the meat for those who like a very hot dish.

SHIVAJI RAO & DEVI HOLKAR
COOKING OF THE MAHARAJAS

Beans, lentils and peas need boiling to destroy toxins (box, page 15).

Yogurt, Grains and Fruit

Muesli

To serve 4

¼ cup	wheat berries, soaked for at least 12 hours and drained	50 ml.
¼ cup	rolled oats, soaked for 30 minutes and drained	50 ml.
1 cup	yogurt	¼ liter
3 tbsp.	honey	45 ml.
1 tsp.	lemon juice	5 ml.
2	medium-sized apples, quartered, cored and grated, peels included	2
1	banana, sliced	1
1 cup	fresh berries or seedless grapes	¼ liter

In a large bowl, combine the yogurt, honey and lemon juice. Add the apples and banana, stirring to mix in the fruits and to keep them from discoloring. Stir in the drained wheat and oats, add the berries or grapes, stir, and serve immediately.

NANCY ALBRIGHT
RODALE'S NATURALLY GREAT FOODS COOKBOOK

Mr. Herbert Bower's Frumenty

Frumenty is perhaps the oldest English national dish. The main ingredient is wheat berries very slowly stewed, or creed, in water. This creed wheat, when cold, is a firm jelly in which the burst grains of wheat are embedded. The wheat is boiled with milk to make frumenty, which may be eaten as a breakfast food with milk and honey or treacle.

To make creed wheat, wash 1 cup [¼ liter] of wheat berries, and put them in a pan or stoneware jar with a lid. Cover the wheat with cold water three times its own measure. Put the wheat into an oven set at its lowest temperature and leave it for at least 12 hours. The grains should have burst and set in a thick jelly. If they have not, boil the contents of the pan for five minutes, or until a jelly is obtained.

To serve 6

2½ cups	creed wheat	625 ml.
5⅓ cups	milk	1⅓ liters
1 tsp.	ground allspice	5 ml.
2 tbsp.	flour	30 ml.

Put the creed wheat and 5 cups [1¼ liters] of the milk into a pan and boil them for 10 to 15 minutes. When the mixture begins to thicken, add the allspice. Finally, stir the remaining milk into the flour to make a thin cream and stir the mixture into the frumenty. Boil the frumenty until it is thick and creamy, and serve.

FLORENCE WHITE (EDITOR)
GOOD THINGS IN ENGLAND

Cracked Wheat and Tomatoes

Burghol ala Banadoora

To serve 4

1 cup	cracked wheat	¼ liter
1	onion, chopped	1
¼ cup	oil	50 ml.
1¼ cups	puréed tomato (recipe, page 166)	300 ml.
	salt and pepper	

Sauté the onion in oil over medium heat until it is translucent—about five minutes. Add the cracked wheat and cook, stirring occasionally, for two minutes. Add 2 cups [½ liter] of boiling water and the puréed tomato. Reduce the heat to low, and cook the mixture until all of the water is absorbed, about 20 minutes. Add salt and pepper to taste.

HELEN COREY
THE ART OF SYRIAN COOKERY

Wheat Pilaf

This pilaf can be baked in a covered casserole in a preheated 350° F. [180° C.] oven for 40 minutes. Then remove the cover, fluff the ingredients, and bake 15 minutes longer.

To serve 4

1 cup	cracked wheat	¼ liter
1	onion, chopped	1
½	green pepper, seeded, deribbed and chopped (optional)	½
1 cup	sliced fresh mushrooms (optional)	¼ liter
¼ cup	olive oil	50 ml.
2 tbsp.	butter	30 ml.
2 cups	chicken or beef stock (recipes, page 166)	½ liter
¼ cup	tomato sauce (recipe, page 167)	50 ml.
	ground thyme	
	turmeric	
	salt	

If you are including the green pepper and mushrooms, place them with the onion in a skillet and sauté them in the oil for five minutes, or until soft. Remove and reserve them. Melt the butter into the oil in the skillet, then add the cracked wheat. Sauté the wheat until it is golden and glossy.

In a saucepan, bring the stock to a boil and add the tomato sauce, a pinch of thyme and turmeric, and salt to taste. Simmer for five minutes, then pour the mixture over the wheat. Cover, and cook over low heat for 30 minutes, or until all the liquid is absorbed. Remove the cover, fluff the wheat with a fork, and dry it over low heat for five minutes.

DOLORES RICCIO & JOAN BINGHAM
MAKE IT YOURSELF

Cracked Wheat with Mushrooms

To serve 6

1½ cups	cracked wheat	375 ml.
1 cup	sliced fresh mushrooms	¼ liter
½ cup	chopped onion	125 ml.
4 tbsp.	butter	60 ml.
3½ cups	chicken stock (recipe, page 166)	875 ml.
1 tsp.	salt	5 ml.
	freshly ground pepper	

Sauté the mushrooms and the onion in the butter, stirring from time to time, for five minutes, or until they are lightly browned. Add the cracked wheat and cook it for five minutes, stirring constantly.

Add the stock, salt and pepper. Bring the mixture to a boil, then reduce the heat, cover the pan, and simmer, stirring occasionally, for 30 minutes, or until all the stock is absorbed and the wheat is soft.

PICTURE COOK BOOK

Bean and Grain Assemblies

Dill and Fava Bean Polo

Sheved Baghala

To serve 8

½ cup	dried fava beans, soaked overnight and drained	125 ml.
2 cups	unprocessed long-grain white rice, soaked for at least 2 hours	½ liter
	salt	
¼ tsp.	saffron threads (optional)	1 ml.
8 tbsp.	butter	120 ml.
1	bunch finely cut fresh dill	1

Peel to remove the beans' tough and bitter skins. Place the beans in a large pot, cover with water, cover the pot and cook the beans for one and one half to two hours, or until they are tender. Drain the beans.

Fill a large saucepan with cold water, place it over high heat and bring to a boil. Add the rice, 2 tablespoons [30 ml.]

of salt and the beans. Boil, uncovered, over high heat, stirring occasionally, for seven to eight minutes. Remove the pan from the heat and drain the rice and beans in a colander. Rinse with lukewarm water.

In the pot in which the rice was cooked, put ¼ cup [50 ml.] of hot water, the saffron (if desired), the butter and ½ teaspoon [2 ml.] of salt; bring the liquid to a boil, then pour half of it into a cup and set it aside.

Mix the dill with the rice and beans, and return the mixture to the pot. Mound the mixture and, with the handle of a spoon, make a deep hole in the center. Cover the pan and cook over medium heat for five minutes; uncover, and sprinkle the rice and beans with the reserved buttery liquid; then cover and steam over very low heat for about 40 minutes.

Remove the pan and place it on a cool surface for about five minutes to loosen the bottom crust. Uncover the pan, and fluff the rice and beans onto a serving dish. Remove the bottom crust and arrange it around the rice and beans or on top of them.

DAISY INY
THE BEST OF BAGHDAD COOKING

Bean and Chicken Stew

The volatile oils in hot chilies may make your skin sting and your eyes burn; after handling chilies, avoid touching your face and wash your hands promptly.

To serve 4 to 6

1 cup	dried pinto beans, soaked for 2 hours	¼ liter
1 cup	dried soybeans, soaked for 2 hours	¼ liter
½ cup	unprocessed white rice	125 ml.
10 oz.	cooked chicken, chopped (about 2 cups [½ liter])	300 g.
½ cup	finely chopped onion	125 ml.
3 or 4	fresh hot green chilies, finely chopped (about ¼ cup [50 ml.])	3 or 4
½	garlic clove, finely chopped	½
1 tsp.	prepared mustard	5 ml.
1 tsp.	salt	5 ml.
2 cups	chicken stock (recipe, page 166)	½ liter

Cook the beans together in the soaking water until soft— one and one half to two hours—adding more water if the beans begin to dry out. Add all of the remaining ingredients and simmer for about 40 minutes before serving.

LYNN B. VILLELLA AND PATRICIA GINS (EDITORS)
GREAT GREEN CHILI COOKING CLASSIC

Beans, lentils and peas need boiling to destroy toxins (box, page 15).

Louisiana Red Beans and Rice

The techniques of cooking rice are explained on pages 22-23. About 4 cups [1 liter] of rice are enough for this dish.

To serve 12

2 cups	dried red beans, soaked overnight and drained	½ liter
	cooked white rice	
1	ham hock	1
1 tbsp.	salt	15 ml.
4	medium-sized onions, chopped	4
6 to 8	scallions, chopped	6 to 8
2	garlic cloves, finely chopped	2
1	green or red pepper, seeded, deribbed and chopped	1
1 cup	finely chopped fresh parsley	¼ liter
1 tsp.	cayenne pepper	5 ml.
1 tsp.	pepper	5 ml.
⅛ tsp.	Tabasco sauce	½ ml.
1 tbsp.	Worcestershire sauce	15 ml.
1 cup	tomato sauce (recipe, page 167)	¼ liter
¼ tsp.	dried oregano leaves	1 ml.
¼ tsp.	dried thyme leaves	1 ml.

Put the beans and ham hock in a large soup kettle, add 2 quarts [2 liters] of water and the salt, and bring them to a boil. Reduce the heat and cook the beans slowly for 45 minutes, skimming occasionally. Add all of the other ingredients except the rice, and continue cooking slowly for two hours, or until the beans are tender and the liquid is thick. Serve over the rice.

THE JUNIOR LEAGUE OF CORPUS CHRISTI
FIESTA: FAVORITE RECIPES OF SOUTH TEXAS

Rice and Peas

To serve 4

1 cup	dried pigeon peas, soaked overnight in 3 cups [¾ liter] water	¼ liter
1 cup	unprocessed white rice	¼ liter
2 cups	coconut milk	½ liter
1 tsp.	dried thyme leaves	5 ml.
1	green pepper, seeded, deribbed and chopped	1

Add the rice, coconut milk, thyme and green pepper to the peas in their soaking water. Bring the liquid to a boil, then cover the pot and cook gently for one and one half to two hours, or until the peas are tender; add more water during cooking if the mixture begins to dry out. Serve hot.

CONNIE AND ARNOLD KROCHMAL
CARIBBEAN COOKING

Black-eyed Peas with Bacon

The techniques of cooking rice are explained on pages 22-23. The volatile oils in hot chilies may make your skin sting and your eyes burn; after handling chilies, avoid touching your face and wash your hands promptly.

To serve 4 to 6

2 cups	dried black-eyed peas, soaked overnight and drained	½ liter
2 cups	cooked white rice	½ liter
2 oz.	lean bacon, 1 oz. [30 g.] cut into matchsticks, the remainder sliced and broiled	60 g.
1	large onion, chopped	1
1	small fresh hot red chili, chopped, or 1 dried hot chili, chopped and soaked in a few spoonfuls of water for 30 minutes	1
2 tbsp.	wine vinegar or cider vinegar	30 ml.
	salt and pepper	
2 tbsp.	chopped fresh parsley	30 ml.

Put the peas in a saucepan. Cover them with salted water and simmer until tender—about one hour. Fry the bacon matchsticks in their own fat until they begin to brown, starting them off in a cold skillet. Drain on paper towels.

In the same skillet, sauté the onion in the bacon fat until tender and very lightly browned. Add the fresh chili or the dried chili and its soaking liquid.

When the black-eyed peas are tender, drain them, reserving ½ cup [125 ml.] of the cooking liquid. Mix together the peas, bacon matchsticks, onion mixture, rice and vinegar. This should be done with a delicate touch to avoid mashing the peas—use a chopstick or the handle of a wooden spoon. Add salt and pepper if necessary.

Turn the mixture into a buttered casserole. Dribble on the reserved cooking liquid, lay the bacon slices on top, and bake in a preheated 350° F. [180° C.] oven for about 20 minutes. Sprinkle with parsley and serve.

MIRIAM UNGERER
GOOD CHEAP FOOD

Hopping John

This unusually named dish from the Deep South is supposed to bring good luck if eaten on New Year's Day before noon.

Sliced sweet onion or pepper relish can be served with Hopping John.

To serve 6

1 cup	dried black-eyed peas, soaked overnight and drained	¼ liter
1 cup	unprocessed white rice	¼ liter
1	ham hock, split	1
1	onion, chopped	1
½ tsp.	crushed dried hot red chili	2 ml.
1 tsp.	salt	5 ml.

Place the ham hock, onion, red chili and enough water to cover them in a saucepan. Bring to a boil and simmer, covered, for 30 minutes. Add the ham-hock mixture with its liquid to the black-eyed peas, adding more water to cover them, if needed. Bring to a boil, skim, and simmer the black-eyed peas for 45 minutes, or until they are tender. Add more boiling water as needed to keep everything moistened.

Add the rice and salt, stir lightly, cover, and simmer for 20 minutes, or until the rice is tender. Add more boiling water if needed. Serve hot.

JEANNE A. VOLTZ
THE FLAVOR OF THE SOUTH

Mixed Beans with Pumpkin

Tbikha à la Courge Rouge

To serve 4

½ cup	dried chick-peas, soaked overnight and drained	125 ml.
½ cup	dried fava beans, soaked overnight, drained and peeled	125 ml.
2	onions, cut into pieces	2
¾ cup	olive oil	175 ml.
1 cup	puréed tomato *(recipe, page 166)*	¼ liter
1½ tsp.	*harissa*	7 ml.
1½ tsp.	cayenne pepper	7 ml.
1 lb.	pumpkin, peeled, seeded and diced	½ kg.
1	medium-sized red pepper, seeded, deribbed and chopped	1
1	bunch fresh parsley	1
	salt	

In a deep casserole, fry the onions gently in the olive oil. Add the puréed tomato, *harissa*, cayenne pepper, chick-peas and

fava beans. Pour in 2 cups [½ liter] of water, cover the pan, and cook over medium heat for two to three hours, or until the beans and peas are tender. Add the pumpkin and the pepper, together with the parsley. Season with salt, and cook the mixture over low heat for a further 30 minutes, until the pumpkin pieces are soft but still whole.

AHMED LAASRI
240 RECETTES DE CUISINE MAROCAINE

Split Peas with Barley

To serve 6

¾ cup	dried split peas	175 ml.
¼ cup	soy grits	50 ml.
¾ cup	pearl barley	175 ml.
2 tbsp.	oil	30 ml.
¼ cup	chopped onion	50 ml.
¼ cup	chopped green pepper	50 ml.
3½ cups	meat stock *(recipe, page 166)*	875 ml.
about ¼ cup	chopped fresh parsley	about 50 ml.
3 tbsp.	finely cut fresh dill	45 ml.
½ tsp.	salt	2 ml.

Heat the oil in a heavy saucepan, and sauté the onion and green pepper for approximately five minutes. Add the split peas, soy grits and barley, and heat, stirring constantly, until they are coated with oil—approximately three minutes. Add the stock, 2 tablespoons [30 ml.] of the parsley, the dill and the salt. Cover the pot and simmer for approximately one and one quarter hours, or until the barley is tender. Serve hot, garnished with the remaining parsley.

NANCY ALBRIGHT
RODALE'S NATURALLY GREAT FOODS COOKBOOK

Beans, lentils and peas need boiling to destroy toxins (box, page 15).

Vegetable and Barley Stew

This hearty stew, which is a meal in itself, is best when cooked in quantity, so if you do not have a 10- to 15-quart [10- to 15-liter] pot, now is the time to buy or borrow one. The finished stew can be refrigerated for one week or frozen and used through the winter months.

To serve 10 to 12

½ cup	dried green split peas	125 ml.
½ cup	dried yellow split peas	125 ml.
½ cup	dried small lima beans	125 ml.
½ cup	pearl barley	125 ml.
2	slices beef shank, 1 inch [2½ cm.] thick	2
3 to 5	meaty beef neck bones	3 to 5
	pepper	
5 quarts	vegetable stock *(recipe, page 166)* or water	5 liters
2	large onions	2
10 to 12	medium-sized tomatoes, peeled	10 to 12
½ lb.	okra, trimmed and cut crosswise into rounds	¼ kg.
3	carrots, sliced	3
2	celery ribs, sliced	2
1 cup	corn kernels	¼ liter
1 cup	green beans	¼ liter
2	medium-sized parsnips, sliced	2
	salt	

Put the shank, neck bones and 1 teaspoon [5 ml.] of pepper into the stock or water. Simmer briskly for one hour, skimming off any foam as it forms. Add the onions and tomatoes. After 30 more minutes, add the peas, lima beans and barley.

Continue cooking the stew for an additional one and one half hours, stirring occasionally. Add the okra, carrots, celery, corn, green beans and parsnips; add salt to taste. Cook for a final 45 minutes to one hour, or until the meat falls from the bones. Skim off excess fat, and pick out any unpalatable gristle before serving or storing.

NORMA JEAN AND CAROLE DARDEN
SPOONBREAD AND STRAWBERRY WINE

"Corn"

"Corn" in Biblical times meant a mixture of lentils, barley, millet, wheat and cumin.

To serve 4 to 6

¼ cup	dried lentils	50 ml.
¼ cup	pearl barley	50 ml.
¼ cup	millet	50 ml.
¼ cup	cracked wheat	50 ml.
1	large onion, chopped	1
6 tbsp.	butter	90 ml.
¼ tsp.	ground cumin	1 ml.
¾ tsp.	salt	4 ml.

In a saucepan set over medium heat, cook the onion in the butter for about five minutes, until golden; add the remaining ingredients plus 2 cups [½ liter] of water. Cover the saucepan with a tight-fitting lid and cook over low heat for 20 minutes; remove the cover and cook for five minutes more. Stir several times during the cooking to prevent sticking. Serve with meat or fowl, with or without gravy.

EILEEN GADEN
BIBLICAL GARDEN COOKERY

Rice with Lentils, Esau-Style

Mujeddrah

To serve 6 to 8

1 cup	dried lentils	¼ liter
1 cup	unprocessed long-grain white rice	¼ liter
2 tsp.	salt	10 ml.
2	large onions, sliced	2
2 tbsp.	oil	30 ml.

Boil the lentils in 2 cups [½ liter] of water with 1 teaspoon [5 ml.] of the salt for 15 to 20 minutes, or until tender. In another pan, bring 2 cups of water to a boil, and add the rice and the remaining salt. Turn off the heat, and let the rice sit until the lentils are ready.

Rinse and drain the lentils and rice. Bring 1½ cups [375 ml.] of water to a boil. Stir in the partly cooked lentils and rice, cover the pan, and simmer the mixture slowly for about 20 minutes, or until the rice is tender.

Sauté the onions in the oil until they soften and turn golden—five to seven minutes. Add the onions and oil to the cooked rice and lentils, and serve.

JOAN NATHAN
THE JEWISH HOLIDAY KITCHEN

Persian Rice with Lentils and Chicken

To serve 4 to 6

1¼ cups	unprocessed long-grain white rice	300 ml.
¼ cup	dried lentils	50 ml.
¼ cup	dried apricots, coarsely chopped	50 ml.
	salt	
2 oz.	almonds, blanched, peeled and sliced (about ½ cup [125 ml.]), or salted pecans, coarsely chopped	60 g.
¾ lb.	unsalted butter	350 g.
1	medium-sized onion, chopped	1
1 tsp.	pepper	5 ml.
3	whole chicken breasts, boned, skinned and cut into ½-inch [1-cm.] cubes	3
1 tbsp.	finely cut fresh dill	15 ml.

Soak the apricots in warm water for 30 minutes before the cooking time, unless they are very soft, in which case soak them only when you begin to cook.

Bring 3 cups [¾ liter] of water to a boil, seasoning it well with 1 tablespoon [15 ml.] of salt. Add the lentils and cook until they are just becoming tender—15 to 20 minutes. Remove the lentils from the heat, pour them into a colander in the sink and keep them warm under warm running water.

Sauté the nuts in 4 tablespoons [60 ml.] of the butter until golden. Remove them from the pan with a slotted spoon. Sauté the onions in the butter remaining in the pan, adding a little more if necessary, until the onions are soft and just beginning to brown lightly. Remove them from the heat and set them aside with the almonds.

Bring 6 cups [1½ liters] of water to a boil and season well with 2 tablespoons [30 ml.] of salt. Sprinkle in the rice. Stir with a wooden spoon until the water begins to boil again. Then boil without stirring for seven to eight minutes, or until the rice is tender but still slightly crunchy. Pour the rice into a colander and rinse with very hot tap water.

Drain both the rice and lentils well, combine them, sprinkle them with pepper, and toss them lightly with two forks so that they are evenly mixed together.

Melt all of the remaining butter. Pour a quarter of it into a heavy 4-quart [4-liter] pot. Sprinkle in the rice-and-lentil mixture very lightly, forming a sort of pyramid on top of the butter. Pour the remaining melted butter over the mixture. Cover the pot with a folded clean cloth and a tight-fitting lid. Place the pot over low heat for about one hour. Remove the lid and let the rice-and-lentil mixture dry out for about five minutes. Taste for salt and sprinkle more on if necessary.

While the rice-and-lentil mixture is cooking, drain the apricots, squeezing out water with your hand. Have the chicken cubes at room temperature. Add the chicken to the rice mixture and use a fork to toss the cubes gently with the top layer of rice and lentils. The heat of the rice mixture will cook the chicken.

As soon as the chicken cubes are mixed into the rice mixture, add the nuts, sautéed onion, apricots and dill. Toss in gently without disturbing the crisp bottom layer of rice and lentils. Serve immediately, scooping up a little of the crisp bottom layer to top each serving.

JOSÉ WILSON (EDITOR)
HOUSE & GARDEN'S PARTY MENU COOKBOOK

Rice and Lentils
Khichhari

The volatile oils in hot chilies may make your skin sting and your eyes burn; after handling chilies, avoid touching your face and wash your hands promptly.

To serve 6

¾ cup	unprocessed white rice	175 ml.
1 cup	dried lentils, mung beans or green split peas, soaked for 1 hour and drained	¼ liter
2 tbsp.	chopped fresh coriander leaves	30 ml.
1 tbsp.	peeled and grated fresh ginger	15 ml.
4	garlic cloves	4
½ tsp.	turmeric	2 ml.
½ tsp.	paprika or cayenne pepper	2 ml.
½ tsp.	*garam masala*	2 ml.
2 tbsp.	*ghee*	30 ml.
1	onion, finely chopped	1
½ tsp.	cumin seeds	2 ml.
2	fresh hot green chilies, stemmed, seeded and finely chopped (optional)	2
2	tomatoes, peeled, seeded and quartered	2
1	potato, diced	1
1 tsp.	salt	5 ml.

In a mortar, pound 1 tablespoon [15 ml.] of the coriander leaves with the ginger, garlic, turmeric, paprika or cayenne pepper, and the *garam masala*. Heat the *ghee* and lightly fry the onion. Stir in the *garam masala* paste along with the cumin seeds and chilies. Fry for a few minutes, and add the rice and the lentils, split peas or mung beans; add the tomatoes and fry for a further five minutes.

Cover the mixture with water. Add the diced potato. Season the mixture with the salt and bring it to a boil. Reduce the heat, cover and simmer for 20 minutes, or until the rice is cooked. Add extra water if necessary to prevent it from drying up. Serve sprinkled with the rest of the coriander leaves.

JACK SANTA MARIA
INDIAN VEGETARIAN COOKERY

Beans, lentils and peas need boiling to destroy toxins (box, page 15).

Rolled Gram and Rice Pancakes

Masala Dosa

The making of these pancakes is shown on pages 48-49; the recipe below was doubled for the demonstration.

To make five 6-inch [15-cm.] pancakes

½ cup	unprocessed white rice, soaked overnight and drained	125 ml.
½ cup	dried split black gram or pink lentils, soaked overnight and drained	125 ml.
1½ tsp.	salt	7 ml.
	baking soda	
1 tbsp.	rice flour	15 ml.

Potato filling

3	medium-sized potatoes, cut into small cubes	3
4	scallions, white parts only, chopped	4
	salt and pepper	
½ tsp.	paprika	2 ml.
¼ tsp.	ground cumin	1 ml.
¼ tsp.	black mustard seeds	1 ml.
1½ tbsp.	butter, melted	22 ml.

Grind the gram or lentils in a food processor until they are extremely fine, and whisk them to a light frothy paste. Grind the rice more coarsely and mix the two. Stir the mixture well and, if necessary, add about ¼ cup [50 ml.] of water to give it a pouring consistency. Add the salt and put the batter in a warm place, about 95° F. [35° C.], for about 12 hours, or until frothy, indicating that it has fermented. Just before cooking, stir in a pinch of baking soda and the rice flour.

To prepare the potato filling, put the filling ingredients in 1 cup [¼ liter] of water and simmer them until the potatoes are soft—15 to 20 minutes. Drain off any excess water.

In a skillet, melt a little butter and cook the potato mixture over high heat until it obtains a dry consistency—about five minutes. Set the mixture aside, keeping it warm.

Lightly butter a heavy skillet and pour in about ¼ cup of the batter. Tilt the pan to spread the batter evenly. Cook the pancake for two minutes, flip it and cook it a further two minutes. Place about 3 tablespoons [45 ml.] of the filling in the center, fold the pancake over and serve it hot. Repeat this process with the remaining batter and filling.

DHARAMJIT SINGH
INDIAN COOKERY

Red Rice

Short- or round-grain rice may be substituted for glutinous rice. Black sesame seeds are a color variety of sesame seeds commonly found at Asian markets.

To serve 8 to 10

5 cups	glutinous rice, rinsed and drained	1¼ liters
⅔ cup	dried adzuki beans	150 ml.
1 tbsp.	salt	15 ml.
3 tbsp.	black sesame seeds, mixed with 1 tsp. [5 ml.] salt	45 ml.

Place the beans in a large pot, and cover them with at least 7 cups [1¾ liters] of water. Without covering the pot, cook the beans for one and one half to two hours, or until soft. Drain the beans, reserving the cooking water to color the rice.

Combine the rice with the beans, 5 cups [1¼ liters] of the bean water and the salt. Cover the pot and cook the beans and rice over low heat for about 30 minutes.

Serve the mixture on plates or in bowls, sprinkled with the salted sesame seeds.

AYA KAGAWA
JAPANESE COOKBOOK

Plymouth Succotash

The cooking of hominy is explained on pages 14-15.

To serve 8

6 cups	cooked cracked hominy, drained	1½ liters
2 cups	dried navy beans, soaked overnight and drained	½ liter
2½ to 3 lb.	chicken, cut into serving pieces	1¼ to 1½ kg.
2½ lb.	corned beef	1¼ kg.
½ lb.	salt pork	¼ kg.
3	potatoes, boiled for 20 minutes and sliced	3
1	small turnip, boiled for 20 minutes and diced	1

Place the beans in a large pot, add enough boiling water to cover the ingredients, and cook them, covered, for two hours, or until they are tender. While the beans are cooking, place the chicken, corned beef and salt pork into a large pot, cover them with 4 quarts [4 liters] of water, and cook them for two to two and one half hours, or until the beef is tender.

Drain the cooked beans and work them through a sieve. Add them to the simmering meat; add the hominy. Stir in the potatoes and turnips.

ANITA D. BROWN, BARBARA BROWN FEHLE, JAMES L. BROWN
AND STEPHEN D. BROWN
THE COLONIAL HERITAGE COOKBOOK

Standard Preparations

Polenta

This recipe is tailored to the finely ground cornmeal generally available in the United States. Classic polenta is made with coarsely ground cornmeal. To prepare classic polenta, add 2 cups [½ liter] of coarsely ground cornmeal to 4 cups [1 liter] of boiling, lightly salted water; follow the directions below but, at the end of 30 minutes, add 1 cup [¼ liter] or more of boiling water—until the desired thickness is achieved—and continue stirring for another hour.

To make about 2 quarts [2 liters] polenta

2 cups	cornmeal	½ liter
1 tbsp.	salt	15 ml.

Bring 2 quarts [2 liters] of water to a vigorous boil over high heat in a large, heavy pot; add the salt, and then gradually sprinkle in the cornmeal while whisking rapidly and continuously. Continue whisking, without stopping, for 20 to 30 minutes, or until the cornmeal tears away from the sides of the pot as you stir. Serve polenta hot, or pour it onto a dampened marble slab or tea towel, let it cool, then cut it into shapes to be sautéed or baked with sauce.

Rice Crusts

To make 1 rice crust

2 cups	unprocessed long-grain white rice	½ liter

In a broad, heavy 3-quart [3-liter] pot, bring the rice and 4 cups [1 liter] of water to a boil over high heat. Stir once or twice to ensure that no rice grains are stuck to the bottom of the pot. Cover the pot and boil the rice for three to four minutes, until the water level has sunk below the rice and small craters appear in the surface of the rice. Reduce the heat to low and simmer, covered, for 15 to 20 minutes, or until all of the water has been absorbed and the rice grains are soft. Spoon out the rice, leaving a shell about ¼ inch [6 mm.] thick on the bottom and sides of the pot. (Reserve the cooked rice for other uses.)

Return the pot to the stove and cook, uncovered, over the lowest possible heat for one hour, or until the crust is crisp and draws away from the sides and bottom of the pot. Let the crust cool to room temperature, then lift it out with your fingers. The crust should stand at least a day before using. It can be stored indefinitely in an airtight container.

Bean Curd

The technique of making bean curd is demonstrated on pages 20-21. *Nigari* (magnesium chloride)—which produces firm, slightly sweet bean curd—is available at health-food stores, as are pressing sacks and settling boxes. For a tart-flavored bean curd, substitute ¼ cup [50 ml.] of lemon or lime juice or 3 tablespoons [45 ml.] of vinegar for the *nigari;* for a mild-flavored, soft bean curd, substitute 2 teaspoons [10 ml.] of Epsom salts (magnesium sulfate) dissolved in 1 cup [¼ liter] of cold water.

To make one 12- to 16-ounce [350- to 500-ml.] cake

1½ cups	soybeans, soaked overnight, drained and rinsed	375 ml.
1½ tsp.	*nigari*, dissolved in 1 cup [¼ liter] water	7 ml.

In a blender or food processor, blend about half of the beans with 2 cups [½ liter] of water. Pour the purée into a heavy pot containing 2 quarts [2 liters] of water. Blend the remaining beans with an additional 2 cups [½ liter] of water and add to the pot.

In a heavy pot, bring the soybean purée to a boil over moderately high heat. Cook for 10 minutes, stirring occasionally with a wooden spoon to prevent sticking. Meanwhile, set a large colander in a pot and line the colander with a double thickness of moistened cheesecloth or with a moistened pressing sack, fitting the mouth of the sack around the rim of the colander.

Ladle the soybean purée into the cheesecloth or pressing sack. Let the purée cool slightly, then press it with the ladle or wooden spoon to force the soy milk through the cloth. When the cloth is cool enough to handle, twist the ends together to enclose the purée. Press and squeeze the cloth to extract all of the soy milk; discard the pulp.

Bring the soy milk to a boil, stirring constantly. Remove the pot from the heat and stir in one third of the *nigari* or other solidifier. Cover the pot and let the mixture sit for three minutes while the first curds form. Remove the lid and sprinkle in half of the remaining solidifier while stirring the top of the liquid gently back and forth with a wooden spoon. Replace the cover and wait a further three minutes. Repeat the process with the remaining solidifier to form fine, white bean curds, which will be floating in an almost clear liquid.

Ladle the bean curds into a settling box lined with a pressing sack or into a cheesecloth-lined mold with drain holes on all sides. Cover the box or mold, weigh it down with a 4-pound [2-kg.] weight and set it aside—a soft bean curd takes about 35 to 40 minutes to set; a firm bean curd takes from one and one half to two hours. Test for doneness by removing the weight and gently prodding the curd to make sure it has reached the desired consistency. Bean curd can be kept in the refrigerator, covered with water, for up to one week; the water should be changed daily.

Meat Stock

This general-purpose, strong stock will keep for up to a week if it is refrigerated in a tightly covered container and boiled for a few minutes every two days. If frozen, the stock will keep for six months.

To make about 3 quarts [3 liters] stock

2 lb.	beef shank	1 kg.
2 lb.	veal shank	1 kg.
2 lb.	chicken backs, necks and wing tips, plus feet if obtainable	1 kg.
1	bouquet garni, including leek and celery	1
1	garlic bulb	1
2	medium-sized onions, 1 stuck with 2 whole cloves	2
4	large carrots	4

Place a metal rack or trivet in the bottom of a large stockpot to prevent the ingredients from sticking. Fit all of the beef, veal and chicken pieces into the pot, and add enough water to cover them by about 2 inches [5 cm.]. Bring slowly to a boil and, with a slotted spoon, skim off the scum that rises. Do not stir, lest you cloud the stock. Keep skimming, occasionally adding a glass of cold water, until no more scum rises—10 to 15 minutes. Add the bouquet garni, garlic, onions and carrots, pushing them down into the liquid so that everything is submerged. Skim again as the liquid returns to a boil. Reduce the heat to very low, cover the pot with the lid ajar and simmer for four to five hours, skimming at intervals. If the meat is to be eaten, remove the veal after one and one half hours, the beef after three hours.

Ladle the stock into a large bowl through a colander lined with a double layer of cheesecloth or muslin. Let the strained stock cool completely, then remove any traces of fat from the surface with a skimmer and a paper towel; if the stock has been refrigerated to cool it, remove the solid fat with a spoon.

Veal stock. Replace the beef and chicken pieces with about 4 pounds [2 kg.] of meaty veal trimmings—neck, shank or rib tips.

Beef stock. Substitute 4 pounds of beef short ribs or chuck or oxtail for the veal shank and chicken pieces, and simmer for five hours.

Chicken and poultry stock. Old hens and roosters yield the richest stock. Omit the beef and veal and use about 5 pounds [2½ kg.] of carcasses, necks, feet, wings, gizzards and hearts, and simmer for two hours. Turkey, duck, goose, or other poultry stock may be made in the same way.

Lamb stock. Use about 6 pounds [3 kg.] of lamb shank, bones and neck in place of the chicken, beef and veal, and simmer for seven to eight hours.

Vegetable Stock

Boil aromatic vegetables—equal amounts of chopped carrot, leek and celery combined with half those amounts of chopped onion and turnip—with a bouquet garni and a crushed garlic clove for 30 minutes in enough lightly salted water to cover the vegetables. Strain the stock and discard the vegetables and bouquet garni. Alternatively, use the water in which any vegetables have been boiled as a stock or as the cooking liquid for producing a vegetable stock.

Fish Stock

To make about 2 quarts [2 liters] stock

2 lb.	fish heads, bones and trimmings, rinsed and broken into convenient sizes	1 kg.
1	onion, sliced	1
1	carrot, sliced	1
1	leek, sliced	1
1	celery rib, diced	1
1	bouquet garni	1
	salt	
2 cups	dry white wine	½ liter

Place the fish, vegetables and herbs in a large pan. Add 2 quarts [2 liters] of water and season lightly with salt. Bring to a boil over low heat. With a large, shallow spoon, skim off the scum that rises to the surface as the liquid reaches a simmer. Keep skimming until no more scum rises, then cover the pan and simmer for 15 minutes. Add the wine and simmer, covered, for another 15 minutes.

Strain the stock through a colander lined with a double thickness of moistened cheesecloth and placed over a deep bowl. If the stock is to be used for a sauce or an aspic, do not press the solids when straining, lest they cloud the liquid.

Puréed Tomato

To make about 1 cup [¼ liter] purée

4	medium-sized tomatoes, peeled, seeded and coarsely chopped	4

Place the tomatoes in a heavy enameled, tin-lined copper, or stainless-steel saucepan. Stirring frequently, cook the tomatoes over low heat for about 10 minutes, or until most of their juices have evaporated and their flesh has been reduced to a thick pulp. Purée the tomatoes through a food mill or sieve into a bowl. Tightly covered and refrigerated, the purée will keep safely for up to five days.

Tomato Sauce

When fresh, ripe summer tomatoes are not available, use 3 cups [¾ liter] of drained canned Italian-style tomatoes.

To make about 1 ½ cups [375 ml.] sauce

4 or 5	very ripe tomatoes, quartered	4 or 5
1	bay leaf	1
1	large sprig thyme	1
	coarse salt	
1	onion, sliced, or 1 garlic clove, crushed (optional)	1
2 tbsp.	butter (optional)	30 ml.
	freshly ground pepper	
1 to 2 tsp.	sugar (optional)	5 to 10 ml.
1 tbsp.	finely chopped fresh parsley	15 ml.
1 tbsp.	coarsely chopped fresh basil leaves	15 ml.

Place the tomatoes in a large enameled or stainless-steel saucepan with the bay leaf, thyme and a pinch of coarse salt. Add the onion or garlic, if desired. Bring the mixture to a boil, crushing the tomatoes lightly with a wooden spoon. Cook, uncovered, over medium heat for 10 minutes, or until the tomatoes have disintegrated into a thick pulp. Remove the bay leaf and thyme.

Tip the tomatoes into a plastic or stainless-steel sieve placed over a bowl and, using a wooden pestle, push them through the sieve. Discard the skins and seeds and return the sieved pulp to the pan. Cook, uncovered, over low heat until the sauce is reduced to the required consistency. If you like, whisk in a little butter to enrich the sauce.

Season with pepper and more salt, if necessary, and sweeten with the sugar if using canned tomatoes. Sprinkle with parsley and basil.

Handmade Mayonnaise

To prevent curdling, the egg yolks, oil, and vinegar or lemon juice should be at room temperature and the oil should be added very gradually at first. The ratio of egg yolks to oil may be varied according to taste. The prepared mayonnaise will keep for several days in a covered container in the refrigerator. Stir it well before use.

To make about 1 ½ cups [375 ml.] mayonnaise

2	egg yolks	2
	salt and white pepper	
2 tsp.	vinegar or strained fresh lemon juice	10 ml.
1 to 1 ½ cups	oil	250 to 375 ml.

Put the egg yolks in a warmed, dry bowl. Season with a little salt and pepper and whisk for about a minute, or until the yolks become slightly paler in color. Add the vinegar or lemon juice and whisk until thoroughly mixed.

Whisking constantly, add the oil, drop by drop to begin with. When the sauce starts to thicken, pour the remaining oil in a thin, steady stream, whisking rhythmically. Add only enough oil to give the mayonnaise a firm but pourable consistency; if it is too thick, add 1 to 2 more teaspoons [5 to 10 ml.] of vinegar or lemon juice, or of warm water.

Herbed mayonnaise. Remove the stems from 8 to 10 spinach leaves, one bunch of watercress, one bunch of parsley, and about six sprigs of tarragon. Blanch the green leaves for 30 seconds, drain them in a strainer, plunge the strainer into cold water to stop the cooking, and pat the leaves dry on a towel. Purée the leaves through a food mill or in a food processor. Add the purée to the egg yolks along with the vinegar or lemon juice.

Chinese White Sauce

This is a sauce used in Chinese "white cooked dishes," as those cooked without soy sauce are called. The preparation of three such dishes is demonstrated on pages 50-53.

To make ¾ to 1 cup [175 to 250 ml.] sauce

¾ cup	chicken stock (recipe, page 166)	175 ml.
1 or 2	garlic cloves, crushed	1 or 2
½ tbsp.	peeled and grated fresh ginger	7 ml.
½ tsp.	salt	2 ml.
½ tsp.	sugar	2 ml.
3 tbsp.	medium-dry sherry	45 ml.
1 tsp.	Chinese sesame-seed oil (optional)	5 ml.
¼ cup	cornstarch, dissolved in ½ cup [125 ml.] water	50 ml.

Combine all of the ingredients except the dissolved cornstarch. Add the resulting sauce to the dish being cooked, let it come to a boil, then stir in the cornstarch, a little at a time, until the desired thickness is reached.

Recipe Index

All recipes in the index that follows are listed by their English titles except in cases where a food of foreign origin, such as cassoulet or polenta, is universally recognized by its source name. Entries also are organized by the bean, lentil, pea and grain ingredients specified in the recipes. Foreign recipes are listed by country or region of origin. Recipe credits appear on pages 174-176.

General Index/ Glossary

Included in this index to the cooking demonstrations are definitions, in italics, of special culinary terms not explained elsewhere in this volume. The Recipe Index begins on page 168.

Recipe Credits

The sources for the recipes in this volume are shown below. Page references in parentheses indicate where the recipes appear in the anthology.

Ackart, Robert, *Cooking in a Casserole.* Copyright © 1967 by Robert Ackart. By permission of Grosset & Dunlap, Inc.(142).

Adam, Hans Karl, *Das Kochbuch aus Schwaben.* © Copyright 1976 by Verlagsteam Wolfgang Hölker. Published by Verlag Wolfgang Hölker, Münster. Translated by permission of Verlag Wolfgang Hölker(112).

Albright, Nancy, *Rodale's Naturally Great Foods Cookbook.* Copyright © 1977 by Rodale Press, Inc. Permission granted by Rodale Press, Inc., Emmaus, Pa.(103, 158, 161).

Alston, Elizabeth, *The Best of Natural Eating Around the World.* Copyright © 1973 by Elizabeth Alston. Published by the David McKay Company, Inc., New York. By permission of the David McKay Company, Inc.(156).

American Heritage, The editors of, *The American Heritage Cookbook and Illustrated History of American Eating.* Copyright © 1964 by American Heritage Publishing Co., Inc. By permission of the publisher(106, 134).

Anderson, Beth, *Wild Rice for All Seasons Cookbook.* © 1977 Minnehaha Publishing. Published by Minnehaha Publishing, Minnesota. By permission of Beth Anderson and Minnehaha Publishing(143, 145).

Artusi, Pellegrino, *La Scienza in Cucina e l'Arte di Mangiar Bene.* Copyright © 1970 Giulio Einaudi Editore S.p.A., Torino. Published by Giulio Einaudi Editore S.p.A.(139).

Aureden, Lilo, *Was Männern so Gut Schmeckt.* Copyright 1954 Paul List Verlag, München. Published by Paul List Verlag, Munich. Translated by permission of Paul List Verlag(122, 132).

Ayensu, Dinah Ameley, *The Art of West African Cooking.* Copyright © 1972 by Dinah Ameley Ayensu. Published by Doubleday & Company, Inc. By permission of the author(118).

Ayrton, Elisabeth, *The Cookery of England.* Copyright © Elisabeth Ayrton, 1974. Published in 1977 by Penguin Books Ltd., London. By permission of Penguin Books Ltd.(130).

Barry, Naomi and Beppe Bellini, *Food alla Florentine.* Copyright © 1972 by Doubleday & Company, Inc. Published by Doubleday & Company, Inc., New York. Reprinted by permission of the author(100).

Beard, James A., *James Beard's American Cookery.* Copyright © 1972 by James A. Beard. First published by Little Brown and Company, Boston. Published in 1974 by Hart-Davis MacGibbon Ltd., Granada Publishing Ltd., Hertfordshire. By permission of Granada Publishing Ltd.(146). *James Beard's Fowl & Game Bird Cookery.* Copyright © 1979 by James Beard. Reprinted by permission of Harcourt Brace Jovanovich, Inc.(107).

Beck, Simone, *New Menus from Simca's Cuisine.* Copyright © 1979, 1978 by Simone Beck and Michael James. Reprinted by permission of Harcourt Brace Jovanovich, Inc.(130).

Beer, Gretel, *Austrian Cooking and Baking.* Copyright © 1954 by Gretel Beer. Published by Dover Publications, Inc. By permission of John Farquharson Ltd., London(119).

Bennani-Smires, Latifa, *Moroccan Cooking.* Published by Société d'Édition et de Diffusion, Al Madariss—Casablanca(111).

Benton, Peggie, *Finnish Food for your Table.* Copyright © 1960 by Peggie Benton. Published by Bruno Cassirer, Oxford. By permission of the publisher(156).

Booth, George C., *The Food and Drink of Mexico.* Copyright © 1964 by George C. Booth. Published in 1976 by Dover Publications, Inc., New York. By permission of Dover Publications, Inc.(107).

Borer, Eva Maria, *Tante Heidi's Swiss Kitchen.* Copyright © 1965 by Nicholas Kaye Ltd. Text © 1981 by Kaye & Ward Ltd. Published by Kaye & Ward Ltd. By permission of the publisher(153).

Boxer, Arabella, *Nature's Harvest: The Vegetable Cookbook.* Copyright © 1974 by Arabella Boxer. By permission of Contemporary Books, Inc., Chicago(112).

Brown, Anita D. and Barbara Brown Fehle, James L. Brown, Stephen D. Brown, *The Colonial Heritage Cookbook.* Copyright © 1975 by Little Brown House Publishing Company. By permission of Little Brown House Publishing Company, Harpers Ferry, W. Va.(164).

Brown, Elizabeth Burton, *Grains: An Illustrated History with Recipes.* Copyright by Prentice-Hall, Inc., Englewood Cliffs, N.J. By permission of the publishers(145, 153).

Bugialli, Giuliano, *The Fine Art of Italian Cooking.* Copyright © 1977 by Giuliano Bugialli. By permission of Times Books, a division of Quadrangle/The New York Times Book Co., Inc.(137).

Bumgarner, Marlene Anne, *The Book of Whole Grains.* Copyright © 1976 by Marlene Anne Bumgarner. By permission of St. Martin's Press(152, 154).

Candler, Teresa Gilardi, *The Northern Italian Cookbook.* Copyright © 1977 by Teresa Gilardi Candler. Published by McGraw-Hill Book Company. By permission of the publisher(151, 156).

Carreras, Marie-Thérèse and Georges Lafforgue, *Les Bonnes Recettes du Pays Catalan.* © Presses de la Renaissance, 1979. Published by Presses de la Renaissance, Paris. Translated by permission of Presses de la Renaissance(124).

Chamberlain, Narcisse and Narcissa G. Chamberlain, *The Chamberlain Sampler of American Cooking.* Copyright © 1961 by Hastings House, Publishers, Inc. By permission of Hastings House, Publishers, Inc.(92, 144, 152).

Chang, Wonona W. and Irving B. and Helene W. and Austin H. Kutscher, *An Encyclopedia of Chinese Food and Cooking.* Copyright © 1970 by Wonona W. Chang and Austin H. Kutscher. By permission of Crown Publishers, Inc.(110).

Chantiles, Vilma Liacouras, *The Food of Greece.* Copyright © 1975 by Vilma Liacouras Chantiles. Published by Atheneum, New York in 1975. By permission of Vilma Liacouras Chantiles(127, 132).

Clark, Morton G., *French-American Cooking.* Copyright © 1967 by Morton G. Clark (J. B. Lippincott). By permission of Harper & Row, Inc.(103, 151).

Clayton, Bernard, Jr., *The Complete Book of Breads.* Copyright © 1973 by Bernard Clayton Jr. Published by Simon & Schuster, a division of Gulf & Western Corporation. By permission of the publisher(150).

Conran, Terence and Caroline, *The Cook Book.* Copyright © 1980 by Mitchell Beazley Publishers, Ltd. Text © 1980 by Conran Ink Limited. By permission of Mitchell Beazley Publishers, Ltd.(154).

Corbitt, Helen, *Helen Corbitt's Cook Book.* Copyright © 1957 by Helen Corbitt. Reprinted by permission of Houghton Mifflin Company(145).

Le Cordon Bleu. Published by Le Cordon Bleu de Paris, 1932. Translated by permission of Le Cordon Bleu de Paris(93).

Corey, Helen, *The Art of Syrian Cookery.* Copyright © 1962 by Helen Corey. By permission of Doubleday & Company, Inc.(94, 158).

Cox, J. Stevens (Editor), *Guernsey Dishes of Bygone Days.* © James and Gregory Stevens Cox, The Toucan Press, Guernsey, 1974. Published by The Toucan Press, Guernsey. By permission of Gregory Stevens Cox, The Toucan Press(100).

Cutler, Carol, *Haute Cuisine for Your Heart's Delight.* Copyright © 1973 by Carol Cutler. By permission of Clarkson N. Potter, Inc.(113). *The Six-Minute Soufflé and Other Culinary Delights.* Copyright © 1976 by Carol Cutler. By permission of Clarkson N. Potter, Inc.(103).

Darden, Norma Jean and Carole, *Spoonbread and Strawberry Wine.* Copyright © 1978 by Norma Jean Darden and Carole Darden. By permission of Doubleday & Company, Inc.(162).

David, Elizabeth, *Italian Food.* Copyright © Elizabeth David 1954, 1963, 1969. Published by Penguin Books Ltd., London. By permission of Penguin Books Ltd (137).

Davidson, Alan, *North Atlantic Seafood.* Copyright © 1979 by Alan Davidson. Reprinted by permission of Viking Penguin, Inc.(138). *Seafood of South-East Asia.* © Alan Davidson 1976. First published by the author in 1976 at World's End, Chelsea, London. Also published by Macmillan London Ltd. in 1978. By permission of Alan Davidson(133).

Day, Irene F., *Kitchen in the Kasbah.* Copyright © 1975 by Irene F. Day. Published in 1976 by André Deutsch Limited, London. By permission of André Deutsch Limited(157).

de Gouy, Louis P., *The Gold Cook Book.* Copyright © 1948, 1969 by the author. Published by Chilton Book Co. Reprinted by permission of the publisher, Chilton Book Co., Radnor, Pa.(152).

de la Rosa, Angeles and C. Gandia de Fernandez, *Flavors of Mexico.* Copyright © 1978 by 101 Productions, San Francisco. By permission of 101 Productions(155).

Delfs, Robert A., *The Good Food of Szechwan.* Copyright © in Japan 1974 by Kodansha International Ltd. By permission of Kodansha International Ltd.(109).

Deutrom, Hilda (Editor), *Ceylon Daily News Cookery Book.* Published by Lake House Investments Limited, Publishers, Colombo. By permission of the publisher(132).

Devi, E. Maheswari, *Handy Rice Recipes.* © Copyright MPH Publications Sdn Bhd 1971. Published by MPH Publications Sdn Bhd, Singapore, 1971. By permission of MPH Publications Sdn Bhd(126).

Donati, Stella, *Le Famose Economiche Ricette di Petronilla.* © Casa Editrice Sonzogno 1974. Published by Casa Editrice Sonzogno S.p.A., Milan, 1974. Translated by permission of Casa Editrice Sonzogno(139).

Duff, Gail, *Gail Duff's Vegetarian Cookbook.* Copyright © 1978 by Gail Duff. Published by Macmillan, London Ltd. By permission of the publisher(103, 105, 118).

Emerling, Carol G. and Eugene O. Jonckers, *The Allergy Cookbook.* Copyright © 1969 by Carol G. Emerling. Reprinted by permission of Doubleday & Company, Inc.(150).

Eren, Neşet, *The Art of Turkish Cooking.* Copyright © 1969 by Neşet Eren. Published by Doubleday & Company, Inc., Garden City, New York. By permission of Neşet Eren(148).

Feng, Doreen Yen Hung, *The Joy of Chinese Cooking.* First published by Faber and Faber Limited, London in 1952. By permission of Faber and Faber Limited(134).

Foods of the World, *American Cooking, Southern Style; The Cooking of China; Latin-American Cooking; Russian Cooking.* © 1971 by Time Inc.; copyright 1968 by Time Inc.; copyright 1968 by Time-Life Books Inc.; copyright 1969 by Time-Life Books Inc. Published by Time-Life Books, Alexandria(123; 141; 90; 148).

Gaden, Eileen, *Biblical Garden Cookery.* Copyright © 1976 by Eileen Gaden. Published by Christian Herald Books. By permission of the publisher(153, 162).

Georgievsky, N. I. and M. E. Melman, E. A. Shadura, A. S. Shemjakinsky, *Ukrainian Cuisine.* © English translation, Technika Publishers, 1975. Published by Technika Publishers, Kiev, 1975. By permission of VAAP, The Copyright Agency of the USSR, Moscow(153).

Giusti-Lanham, Hedy and Andrea Dodi, *The Cuisine of Venice & Surrounding Northern Regions.* Copyright © 1978 by Barron's Educational Series Inc. Reprinted by permission of Barron's(140, 150).

Gorman, Marion and Felipe P. de Alba, *The Dione Lucas Book of Natural French Cooking.* Copyright © 1977 by Marion Gorman and Felipe P. de Alba. Published by E. P. Dutton. Reprinted by permission of Hawthorn Properties (Elsevier-Dutton Publishing Co., Inc.)(93).

Grigson, Jane, *English Food.* Copyright © Jane Grigson, 1974. First published by Macmillan 1974. Published by Penguin Books 1977. By permission of Macmillan, London and Basingstoke(120).

Guasch, Juan Castelló, *Bon Profit! El Libro de la Cocina Ibicenca.* Published in 1967 by Imprenta Alfa, Palma, Majorca. Translated by permission of Juan Castelló Guasch(123).

Gupta, Pranati Sen, *The Art of Indian Cuisine.* Copyright © 1974 by Pranati Sen Gupta. Reprinted by permission of the publisher, E. P. Dutton. (A Hawthorn Book)(128).

Hawkes, Alex D., *A World of Vegetable Cookery.* Copyright © 1968 by Alex D. Hawkes. Reprinted by permission of Simon & Schuster, a division of Gulf & Western Corporation(117).

Hawliczkowa, Helena, *Kuchnia Polska.* (Editor: Maria Librowska.) Published by Panstwowe Wydawnictwo Ekonomiczne, Warsaw, 1976. Translated by permission of Agencja Autorska, for the author(147).

Hazelton, Nika, *The Regional Italian Kitchen.* Copyright © 1978 by Nika Hazelton. Published by M. Evans and Company, Inc., New York. By permission of Curtis Brown Ltd., New York(131).

Heller, Edna Eby, *The Art of Pennsylvania Dutch Cooking.* Copyright © 1968 by Edna Eby Heller. Reprinted by permission of Doubleday & Company, Inc.(149).

Henderson, Mary, *Paris Embassy Cookbook.* Copyright © 1980 by Mary Henderson. Published by Weidenfeld & Nicholson, London. By permission of the publisher(96).

Hewitt, Jean, *The New York Times Weekend Cookbook.* Copyright © 1975 by Jean Hewitt. By permission of Times Books, a division of Quadrangle/The New York Times Book Co., Inc.(113).

Historia Hors Série, No. 42: Les Français à Table (bimonthly review). © Librairie Jules Tallandier, 1975. Published by Librairie Jules Tallandier, Paris. Translated by permission of Librairie Jules Tallandier(99).

Holkar, Shivaji Rao & Shalini Devi, *Cooking of the Maharajas.* Copyright © 1975 by Shivaji Rao Holkar and Shalini Devi Holkar. First published in 1975 by The Viking Press, Inc., New York. By permission of Shivaji Rao Holkar(157).

Hsiung, Deh-Ta, *Chinese Regional Cooking.* Copyright © 1979 by Quarto Publishing Limited. Published by Macdonald Educational Limited, London. By permission of Quarto Publishing Limited(108, 110).

Iny, Daisy, *The Best of Baghdad Cooking.* Copyright © 1976 by Daisy Iny. Published by Saturday Review Press/E. P. Dutton & Co., Inc., New York. By permission of Manuscripts Unlimited, New York for the author(159).

Isnard, Léon, *La Cuisine Française et Africaine.* © Albin Michel, 1949. Published by Éditions Albin Michel, Paris. Translated by permission of Éditions Albin Michel(101).

Jensen, Bodil (Editor), *Take a Silver Dish.* Copyright © 1962 by Høst & Søns Forlag. Published by Høst & Søns Forlag, Denmark(146).

Jervey, Phyllis, *Rice & Spice: Rice Recipes from East to West.* Copyright © 1957 by Charles E. Tuttle Co., Inc. Published by Charles E. Tuttle, Inc., Tokyo. By permission of Charles E. Tuttle Co., Inc.(142).

The Junior League of the City of New York, *New York Entertains.* Copyright © 1974 by the Junior League of New York, Inc. Reprinted by permission of Doubleday & Co., Inc.(143).

The Junior League of Corpus Christi, *Fiesta: Favorite Recipes of South Texas.* Copyright © 1973 by The Junior League of Corpus Christi, Inc. By permission of the publisher, The Junior League of Corpus Christi(160—Mrs. Lev H. Prichard III (Ella Wall)).

The Junior League of New Orleans, *The Plantation Cookbook.* Copyright © 1972 by The Junior League of New Orleans, Inc. Reprinted by permission of Doubleday & Company, Inc.(123, 151).

Kagawa, Aya, *Japanese Cookbook.* © 1967 by Japan Travel Bureau, Inc. Published by Japan Travel Bureau, Inc., Publishing Division, Tokyo. By permission of Japan Travel Bureau, Inc.(164).

Karsenty, Irène and Lucienne, *La Cuisine Pied-Noir (Cuisines du Terroir).* © 1974, by Éditions Denoël, Paris. Published by Éditions Denoël. Translated by permission of Éditions Denoël(115, 116).

Katzen, Mollie, *The Moosewood Cookbook.* Copyright © 1977 Mollie Katzen. Published by Ten Speed Press, Berkeley. By permission of the author(108).

Kennedy, Diana, *The Cuisines of Mexico.* Copyright © 1972 by Diana Kennedy. By permission of Harper & Row,

Khayat, Marie Karam and Margaret Clark Keatinge, *Food from the Arab World.* Copyright © 1959 by Marie Karam Khayat and Margaret Clark Keatinge. Published in 1959, 1961, 1965 by Khayats, Beirut(114).

Koock, Mary Faulk, *The Texas Cookbook.* Copyright © 1965 by Mary Faulk Koock and Rosalind Cole. Reprinted by permission of Little, Brown and Company(118).

Krochmal, Connie and Arnold, *Caribbean Cooking.* Copyright © 1974 by Connie and Arnold Krochmal. By permission of the publisher, Times Books, a division of Quadrangle/The New York Times Book Co., Inc.(120, 122, 149, 160).

Kulinarische Gerichte: Zu Gast bei Freunden. Copyright to this translation by Verlag für die Frau DDR, Leipzig. Published by Verlag für die Frau, Leipzig and Verlag MIR, Moscow, 1977. Translated by permission of VAAP, The Copyright Agency of the USSR, Moscow(113).

Kürtz, Jutta, *Das Kochbuch aus Schleswig-Holstein.* © Copyright 1976 by Verlagsteam Wolfgang Hölker. Published by Verlag Wolfgang Hölker, Münster. Translated by permission of Verlag Wolfgang Hölker(145).

Laasri, Ahmed, *240 Recettes de Cuisine Marocaine.* © 1978, Jacques Grancher, Éditeur. Published by Jacques Grancher, Éditor, Paris. Translated by permission of Jacques Grancher, Éditeur, Paris(101, 161).

Lane, Lilian, *Malayan Cookery Recipes.* Copyright © 1964 by Lilian Lane. First published in Singapore in 1964 by Eastern Universities Press Ltd. for University of London Press Ltd. By permission of University of London Press Ltd.(134).

Lang, George, *The Cuisine of Hungary.* Copyright © 1971 by George Lang (New York: Atheneum, 1971). Reprinted with the permission of Atheneum Publishers(93).

Lin, Florence, *Florence Lin's Chinese One-Dish Meals.* Copyright © 1978 by Florence Shen Lin. Reprinted by permission of the publisher, E. P. Dutton. (A Hawthorn Book)(136). *Florence Lin's Chinese Regional Cookbook.* Copyright © 1975 by Florence Lin. Reprinted by permission of the publisher, E. P. Dutton. (A Hawthorn Book)(135).

Lopez, Candido, *El Libro de Oro de la Gastronomia.* © 1979 by Candido Lopez. Published by Plaza & Janes S.A., Barcelona. Translated by permission of Plaza & Janes S.A.(124).

Majumder, P., *Cook Indian.* Copyright © 1980 by Times Books International. By permission of Times Books International, Singapore(94, 120).

Mallos, Tess, *The Complete Middle East Cookbook.* Copyright © 1979 by Tess Mallos. By permission of the publisher, McGraw-Hill Book Company, New York(133).

Manual de Cocina. Published by Editorial Almena, Instituto del Bienestar, Ministerio de Cultura, Madrid, 1965. Translated by permission of Editorial Almena, Instituto del Bienestar, Ministerio de Cultura(131).

Mardikian, George, *Dinner at Omar Khayyam's.* Copyright © 1945 by George Mardikian. Copyright renewed. Published by The Viking Press, New York. By permission of McIntosh and Otis, Inc., New York(126).

Maria, Jack Santa, *Indian Vegetarian Cookery.* © Jack Santa Maria 1973. Published by Rider and Company, an imprint of the Hutchinson Publishing Group, London. By permission of the Hutchinson Publishing Group(115, 120, 163).

Marty, Albin, *Fourmiguetto: Souvenirs, Contes et Recettes du Languedoc.* Published by Éditions CREER, Nonette, 1978. Translated by permission of Éditions CREER(117).

Meade, Martha, *Recipes from the Old South.* Copyright © 1961 by Martha Meade. Reprinted by permission of Holt, Rinehart and Winston, Publishers(143, 144).

Mei, Fu Pei, *Pei Mei's Chinese Cook Book, Vol. II.* Published by T & S Industrial Co., Ltd., Taipei. By permission of the author(141).

Miller, Amy Bess and Persis W. Fuller (Editors), *The Best of Shaker Cooking.* Copyright © 1970 by Shaker Community, Inc. Reprinted by permission of Macmillan Publishing Co., Inc.(149).

Mitchell, Alice Miller (Editor), *Oriental Cookbook.* Copyright 1954 by Alice Miller Mitchell. Published by Rand McNally & Company, New York. By permission of Rand

McNally & Company(105).

Il Mondo in Cucina: Minestre, Zuppe, Riso. Copyright 1969, 1971 by Time Inc. Jointly published by Sansoni/Time-Life(138).

Nathan, Joan, *The Jewish Holiday Kitchen.* Copyright © 1979 by Schocken Books Inc. Reprinted by permission of Schocken Books Inc.(162).

Nignon, M. Édouard (Editor), *Le Livre de Cuisine de L'Ouest-Éclair.* Published in 1941 by l'Ouest-Éclair, Rennes. Translated by permission of Société d'Éditions Ouest-France(95, 97).

Norberg, Inga, *Good Food from Sweden.* Copyright © 1939 by Inga Norberg. Published by William Morrow & Co. Originally published by Barrows. By permission of William Morrow & Co.(129).

Norman, Barbara, *The Spanish Cookbook.* Copyright © 1966 by Barbara Norman. Published by Atheneum. By permission of the author(97).

Ochorowicz-Montowa, Marja, *Polish Cooking.* (Translated and adapted by Jean Karsavina.) Copyright © 1958 by Crown Publishers, Inc. By permission of Crown Publishers, Inc.(121, 135, 147).

Olney, Richard, *The French Menu Cookbook.* Copyright © 1970 by Richard Olney. Published by Simon and Schuster, New York. By permission of John Schaffner, Literary Agent, New York(98).

The Original Picayune Creole Cook Book. Copyright © 1901, 1906, 1916, 1922, 1928, and 1936 by The Times-Picayune Publishing Company. Published by The Times-Picayune Publishing Company. By permission of the publisher(104).

Paddleford, Clementine, *The Best in American Cooking.* Copyright © 1960 Clementine Paddleford. Copyright © 1970 Chase Manhattan Bank, Executors of the Estate of Clementine Paddleford. Published by Charles Scribner's Sons, New York. By permission of Charles Scribner's Sons(106).

Perl, Lila, *Rice, Spice and Bitter Oranges.* Text copyright © 1967 by Lila Perl. Published by The World Publishing Company, Cleveland. By permission of the author(122).

Philpot, Rosl, *Viennese Cookery.* Copyright © 1965 by Rosl Philpot. By permission of the publisher, Hodder and Stoughton, Ltd., London(119).

Picture Cook Book. Copyright © 1958, 1968 by Time Inc. Published by Time-Life Books, Alexandria(148, 159).

Pohren, Donn E., *Adventures in Taste: The Wines and Folk Food of Spain.* Copyright © 1972 by Donn E. Pohren. By permission of the author(111, 116).

Puga Y Parga, Manuel M. (Picadillo), *La Cocina Práctica.* Fifth edition. Privately published in La Coruña(112).

Reich, Lilly Joss, *The Viennese Pastry Cookbook.* Copyright © 1970 by Lilly Joss Reich. Published by Macmillan Publishing Co., Inc. By permission of the publisher(106).

Restino, Susan, *Mrs. Restino's Country Kitchen.* Copyright © 1976 by Susan Restino. Published by Quick Fox, New York. By permission of the publisher(90).

Riccio, Dolores & Joan Bingham, *Make It Yourself.* Copyright © 1978 by Dolores Riccio and Joan Bingham. Published by Chilton Book Company, Radnor, Pa. By permission of the publisher(158).

Riker, Tom & Richard Roberts, *The Directory of Natural and Health Foods.* Copyright © 1979 by Tom Riker & Richard Roberts. Reprinted by permission of G. P. Putnam's Sons(108, 142).

Sahni, Julie, *Classic Indian Cooking.* Copyright © 1980 by Julie Sahni. By permission of William Morrow & Company(95, 114).

Schapira, Christiane, *La Cuisine Corse.* © Solar, 1979. Published by Solar, Paris. Translated by permission of Solar(156).

Schuler, Elizabeth, *Mein Kochbuch.* © Copyright 1948 by Schuler-Verlag, Stuttgart-N, Lenzhalde 28. Published by Schuler Verlagsgesellschaft. Translated by permission of Schuler Verlagsgesellschaft(105).

Serra, Victoria, *Tia Victoria's Spanish Kitchen.* English text copyright © Elizabeth Gili, 1963. Published by Kaye & Ward Ltd., London, 1963. Translated by Elizabeth Gili from the original Spanish entitled *Sabores: Cocina del Hogar* by

Victoria Serra Suñol. By permission of Kaye & Ward Ltd.(125).
The Settlement Cookbook. Copyright © 1965, 1976 by The Settlement Cookbook Co. Reprinted by permission of Simon & Schuster, a division of Gulf & Western Corporation(130).
Shulman, Martha Rose, *The Vegetarian Feast.* Copyright © 1979 by Martha Rose Shulman. By permission of Harper & Row, Publishers, Inc.(100).
Shurtleff, William & Akiko Aoyagi, *The Book of Tofu.* Copyright © 1975 by William Shurtleff and Akiko Aoyagi. By permission of the publisher, Autumn Press, Brookline, Massachusetts(109, 110).
Singh, Dharamjit, *Indian Cookery.* Copyright © 1970 by Dharamjit Singh. By permission of the publisher, Penguin Books(127, 164).
Solomon, Charmaine, *The Complete Asian Cookbook.* Copyright © 1976 by Paul Hamlyn Pty Limited. By permission of Lansdowne Press, Sydney(94, 140).
Stavonhagen, Mary Ma, *The Complete Encyclopedia of Chinese Cooking.* Copyright Octopus Books Limited, London. By permission of Octopus Books Limited(136).
Tarr, Yvonne Young, *The Farmhouse Cookbook.* Copyright ©1973 by Yvonne Young Tarr. By permission of Times Books, a division of Quadrangle/The New York Times Book Co., Inc.(149).
Tharp, Edgar and Robert E. Jaycoxe, *The Starving Artist's Cookbook.* Copyright © 1976 by Edgar Tharp & Robert E. Jaycoxe. By permission of the publisher, McGraw-Hill Book Company(152).
Tracy, Mirian, *Real Food: Simple, Sensuous & Splendid.* Copyright 1978 by Mirian Tracy. Reprinted by permission of Viking Penguin Inc.(92).
Troisgros, Jean and Pierre, *The Nouvelle Cuisine of Jean & Pierre Troisgros.* Copyright © 1978 in the English

translation by William Morrow and Company, Inc. Originally published under the title of *Cuisiniers à Roanne* © 1977 by Éditions Robert Laffont, S.A. By permission of William Morrow and Company, Inc.(96).
Turgeon, Charlotte and Frederic A. Birmingham, The Saturday Evening Post All American Cookbook. Copyright © 1976 by The Curtis Publishing Company, U.S.A. First published in Great Britain in 1977 by Elm Tree Books Ltd., Hamish Hamilton Ltd., London. By permission of Hamish Hamilton Ltd.(105).
Tzabar, Naomi & Shimon, *Yemenite & Sabra Cookery.* Copyright © 1963, 1966, 1974 and SADAN Publishing House Ltd. Published by SADAN Publishing House Ltd., Tel-Aviv. By permission of SADAN Publishing House Ltd.(117).
Ungerer, Miriam, *Good Cheap Food.* Copyright © 1973 by Miriam Ungerer. Reprinted by permission of Viking Penguin Inc.(104, 160).
Uvezian, Sonia, *The Best Foods of Russia.* Copyright © 1976 by Sonia Uvezian. Reprinted by permission of Harcourt Brace Jovanovich, Inc.(102).
Valldejuli, Carmen Aboy, *Puerto Rican Cookery.* Copyright © 1977 by Carmen Aboy Valldejuli. Privately published in Santurce. By permission of the author(101).
Verdon, René, *The White House Chef Cookbook.* Copyright © 1967 by René Verdon. Reprinted by permission of Doubleday & Company, Inc.(121).
Vidal, Coloma Abrinas, *Cocina Selecta Mallorquina.* Published in 1975 by Imprenta Roig, Campos, Majorca. Translated by permission of Señora Suñer, for the author(118).
Villella, Lynn B. and Patricia Gins (Editors), *Great Green Chili Cooking Classic.* Copyright © 1974 by The Albuquerque Tribune. By permission of the publisher, The Albuquerque Tribune(159).

Voltz, Jeanne A., *The Flavor of the South.* Copyright © 1977 by Jeanne A. Voltz. Reprinted by permission of Doubleday & Company, Inc.(125, 150, 161).
White, Florence (Editor), *Good Things in England.* Published by arrangement with Jonathan Cape Ltd., 1968, by The Cookery Book Club, London. By permission of Jonathan Cape Ltd.(158).
Willan, Anne, *French Cookery School.* Copyright © 1980 by Anne Willan and Jane Grigson. Published by Macdonald Futura Publishers. Reprinted by permission of La Varenne, Paris(128).
Willinsky, Grete, *Kochbuch der Büchergilde.* © Büchergilde Gutenberg, Frankfurt am Main 1958. Published by Büchergilde Gutenberg, 1967. Translated by permission of Büchergilde Gutenberg(119).
Wilson, José (Editor), *House & Garden's Party Menu Cookbook.* Copyright © 1973 by The Condé Nast Publications Inc. Published by The Condé Nast Publications Inc., New York. By permission of The Condé Nast Publications Inc.(128, 144, 163).
The Wise Encyclopedia of Cookery. Copyright © 1948 by Wm. H. Wise & Co., Inc. By permission of Grosset & Dunlap, New York(142).
Wolfert, Paula, *Couscous and Other Good Food from Morocco.* Copyright © 1973 by Paula Wolfert. Published by Harper & Row Publishers, Inc. By permission of Harper & Row and Paula Wolfert(146).
Yianilos, Theresa Karas, *The Complete Greek Cookbook.* Copyright © 1970 by Theresa Karas Yianilos. By permission of Avenel Books, a division of Barre, by arrangement with Funk and Wagnalls(96, 126).
Zuliani, Mariù Salvatori de, *La Cucina di Versilia e Garfagnana.* Copyright © by Franco Angeli Editore, Milano. Published in 1969 by Franco Angeli Editore, Milan. Translated by permission of Franco Angeli Editore(139).

Acknowledgments

The indexes for this book were prepared by Louise W. Hedberg. The editors are particularly indebted to Stu Berman, The Court of the Mandarins, Washington, D.C.; Gail Duff, Kent, England; Dr. Alfred C. Olson, U.S. Department of Agriculture, Western Regional Center, Albany, California; Ann O'Sullivan, Majorca, Spain; Julie Sahni, Brooklyn Heights, New York; Dr. Hwa L. Wang, U.S. Department of Agriculture, North Regional Research Center, Peoria, Illinois; Dr. B. D. Webb, U.S. Department of Agriculture, Beaumont, Texas.

The editors also wish to thank: Apothecary Shop, Alexandria, Virginia; Caroline Baum, London; Birkett Mills, Penn Yann, New York; Harold Blain, Washington and Idaho Dry Pea and Lentil Commissions, Moscow, Idaho; Eliott Burgess, Majorca, Spain; Jacqueline Cattani, Bethesda, Maryland; Josephine Christian, Bath, England;

Emma Codrington, Surrey, England; Clare Coope, London; Nona Coxhead, London; Jennifer Davidson, Berkeley, California; Fiona Duncan, London; Mimi Errington, London; Boyd Foster, Arrowhead Mills, Inc., Hereford, Texas; Madeline Fox, John Lane, New England Soy Dairy, Springfield, Massachusetts; Geoffrey Grigson, Swindon, England; Dr. C. R. Gunn, Plant Taxonomy Laboratory, U.S. Department of Agriculture, Beltsville, Maryland; Rae Hartfield, Mary Alice Volkert, The Rice Council, Houston, Texas; Maggie Heinz, London; Richard R. Holcomb, Jr., Minnesota Grain Pearling Company, Cannon Falls, Minnesota; Nell Hopson, Uncle Ben's Foods, Houston, Texas; Marion Hunter, Surrey, England; Dora Jonassen, New York, New York; Robert E. Jones, ADM Foods, Shawnee Mission, Kansas; Alison Kerr, Cambridgeshire, England; Rosemary Klein, London; John Lamb, London; Lauhoff Grain Company, Danville, Illinois; Laurelbrook Foods, Raleigh, North Carolina; Dr. I. E. Liener, University of Minnesota, Minneapolis; Florence Lin, Bronx, New York; Lina Stores, London; Dr. Alan Long, The

Vegetarian Society, London; Pippa Millard, London; James Miller, Quong Hop & Company, South San Francisco; Mrs. Miller, Royal Horticultural Society, Wisley, England; Dr. John Montoure, Dr. Karen Davis, University of Idaho, Moscow; Dr. John Moseman, Field Crops Laboratory, U.S. Department of Agriculture, Beltsville, Maryland; Dilys Naylor, Surrey, England; Jo Oxley, Surrey, England; G. Parmigiani, United Preserves Ltd., London; Dr. C. Lorenzo Pope, Rice Researchers, Inc., Glenn, California; Dr. Louis Rockland, Chapman College, Orange, California; Lois Ross, Marcia Watts, The Quaker Oats Co., Chicago, Illinois; Craig Sams, Harmony Foods, London; Michael Schwab, London; William Shurtleff, The Soyfoods Center, Lafayette, California; Dr. J. R. Stavely, U.S. Department of Agriculture, Beltsville, Maryland; Anne Stephenson, London; Fiona Tillett, London; Pat Tookey, London; Dr. Mark Uebersax, Michigan State University, East Lansing; Walnut Acres, Penns Creek, Pennsylvania; Harold West, Idaho Bean Commission, Boise; Paula Wolfert, New York, New York.

Picture Credits

The sources for the pictures in this book are listed below. Credits for each of the photographers and illustrators are listed by page number in sequence with successive pages indicated by hyphens; where necessary, the locations of pictures within pages are also indicated—separated from page number by dashes.

Photographs by Aldo Tutino: cover, 4, 8-11, 16-21, 22— bottom, 23—top right and bottom, 24, 25—left and

center, 27—bottom, 30-41, 44-55, 56—bottom, 58, 60— top and bottom left, 61—top, 62-72, 78-81, 83—bottom right, 86-88. Other photographs (alphabetically): Tom Belshaw, 22— top, 23—top left, 28-29, 42-44, 56—top, 57, 74-77, 82— bottom, 83—top and bottom left, 84—top. John Cook, 84—bottom, 85. David Davies, 60—bottom right, 61— bottom. John Elliott, 12, 25—right, 26, 27—top, 82—top. Louis Klein, 2.
Illustrations: From The Mary Evans Picture Library and private sources and *Food & Drink: A Pictorial Archive from Nineteenth Century Sources* by Jim Harter, published

by Dover Publications, Inc., 1979, 6-7, 93-167.

Library of Congress Cataloguing in Publication Data
Main entry under title:
Dried beans & Grains.
   (The Good cook, techniques & recipes)
   Includes index.
   1. Cookery (Cereals) 2. Cookery (Beans)
I. Time-Life Books. II. Title: Dried Beans and Grains. III. Series.
TX808.D74   641.6'331   81-13578
ISBN 0-8094-2922-5   AACR2
ISBN 0-8094-2921-7 (lib. bdg.)
ISBN 0-8094-2920-9 (retail ed.)